1:1:85

Microwave and Optical Ray Geometry

Microwave and Optical Ray Geometry

S. CORNBLEET

Department of Physics
University of Surrey

A Wiley-Interscience Publication

JOHN WILEY & SONS

Chichester . New York . Brisbane . Toronto . Singapore

Library of Congress Cataloging in Publication Data:

Cornbleet, S.
  Microwave and optical ray geometry.

  'A Wiley-Interscience publication.'
  Includes index.
  1. Optics, Geometrical.   2. Microwave optics.
I. I. Title.
QC381.C827   1984   535'.32   83–16737
ISBN 0 471 90315 9

British Library Cataloguing in Publication Data:

Cornbleet, S.
  Microwave and optical ray geometry.
  1. Microwave optics   2. Optics, Geometrical
  I. Title
  535'.2   QC675.8

  ISBN 0 471 90315 9

Filmset by Mid-County Press, London, SW15
Printed by Pitmen Press, Ltd, Bath, Avon

Contents

vi

Introduction

Geometrical optics has long been considered to be the method of tracing rays through the elements of an optical system, each consisting of a uniform refractive medium, and the determination of the performance of the system from the resulting distribution over the exit pupil. Over recent years, developments have occurred which have enlarged the scope of this method. Originally this came about with the advent of microwave technology, using wide-angle systems with few surfaces but no longer confined to the spherical form. The first microwave antennas were based exclusively on the paraboloid, and currently variations in the design use the same paraboloid in conjunction with a secondary reflector in configurations that are no longer simple and symmetrical. However, the essential geometry based on ray tracing is similar to that of the optical method but the design is usually for as perfectly a focusing system as possible. Also in the field of microwaves has been the use of non-uniform refractive materials although their application has been limited to one or two basic designs. As higher frequencies come into use the microwave designs become more optical in nature, and newer methods are required.

There has been a similar progression in the purely optical field. New in-the-large optical processes are being considered in which whole surface interactions are involved, a phase front with a refracting surface, for example, which can be dealt with without tracing individual rays. Modern technology has also made possible the use of machining techniques for the manufacture of aspheric surfaced components, requiring a new geometrical evaluation. Essentially the design in optics is for many surface systems to obtain as wide a field of view about the nominal focus within given tolerances as is possible. Some applications are more concerned with the light-gathering properties of the optical design and ignore aberrations and focusing altogether.

The recent advent of non-uniform optical fibres has introduced that subject to optics. New techniques, such as ion implantation, make non-uniform optical media possible, and their use in the field of integrated optics requires a knowledge of the ray-tracing techniques and focusing possibilities applicable to that medium.

These advances have shown that a greater appreciation of the fundamental geometry of geometrical optics is required and it is to this end that this book is directed.

The first part deals with reflecting and refracting systems, generally of only a few surfaces and with perfect focusing. It is thus of more interest to the microwave antenna designer but, where extensions to the optical field exist, this has been indicated. The basic laws of refraction and reflection are given in their fundamental differential form in the first chapter and applied to some new combinations of surfaces based on the sphere. It would appear that once a dual or multi-surfaced system is decided upon, there is no commitment to remain confined to the paraboloid as one of the elements concerned. The following chapters lay the foundation for the development of a theory of inversion, a whole surface design technique requiring the definition and geometrical derivation of the concept of the zero-distance phase front. The geometry of this surface is studied in depth and a chapter devoted to the applications of it to perfect focusing and microwave systems. It does not appear to extend easily to wide-field systems, applying at most to multi-focus devices.

There follows a technique for the mechanical drawing of optical surfaces based upon an ancient theorem of Leibnitz regarding caustics and reflectors. This is presented as a string drawing method for ease of conception but is convertible to algebraic geometry. In essence the string itself plays the part of a movable ray, but the geometry involved does qualify as an in-the-large process.

Chapter 5 lays the theoretical foundation for the ray-tracing formulae required for rays in an isotropic non-uniform refractive medium. Apart from the obvious coordinate systems, one or two more general coordinate spaces are examined and the complete ray trace equations derived for them. These are usually too general for complete solution, but when the single condition is made, that the pencil of rays is confined to a single coordinate surface through a symmetry in the medium, solutions are possible for a variety of source or focusing situations.

The following chapters illustrate possible focusing elements in these coordinate systems, again demanding an appreciation of the geometry of unusual curves. In the process of preparing this work a new geometrical property of the ray paths in the non-uniform spherical medium was discovered. This too is an inversion, with a result similar to the inversion in the previous part, that a second design together with its ray trace or phase front, can be obtained directly from a given design, by an inversion transformation. The development of a technique whereby an optical system can be obtained by a transformation from a given optical system, is a long-standing problem in physics. Here we have two such transformations and their practical applications, and both are inversions.

In the final chapter the spherical medium is considered again in a form also applicable to integrated optical components. Rays confined to a curved surface follow the geodesics of the surface, so a correctly formed surface can have focusing or other ray-tracing properties. It is known that the rays in a spherical medium discussed in Chapter 7 are analogous to the geodesics on a surface of revolution and thus the design of the one automatically designs the other.

Geodesics on surfaces of revolution also have an analogy to the trajectories of particles in potential fields and thus so have rays in a spherical medium. The geometrical optics of rays also forms a first-order solution to the electromagnetic field equations. It is of some interest to observe that in the dynamics of particles there has arisen a transformation theory akin to the inversion theory we present here. There exists an inversion of the electromagnetic field which leaves the ray equations invariant. However, it disagrees in some respects with the result in Chapter 3. These are given only a brief discussion in the final chapter, more to indicate their existence than to go into the matter deeply. This will be the subject of other studies.

The appendices list the geometrical properties of the many curves that can arise in the use of the techniques presented, and the fundamental integration process for obtaining the ray path in any of the non-uniform media discussed in the text.

ACKNOWLEDGEMENTS

The author acknowledges with gratitude the assistance of Dr. M. C. Jones of the Department of Physics, The University of Surrey, with many points of clarification included in the text. Permission has been received from Dover Publications for the reproduction of many formulae and the geometry of curves mostly collected in Appendices I and II, and from the editor of the Proceedings of the I.E.E. for the reproduction of parts of articles by the author published in recent years in Part H. Most thanks are due to Mrs. M. Gunney for typing a complicated mathematical script with efficiency and good humour and to the staff of John Wiley & Sons Ltd. for their encouragement and cooperation.

S. Cornbleet
Guildford 1983

1 The Laws of Refraction and Reflection

The term optical geometry is taken to imply the entire body of classical geometry, with the single additional postulation that at some surfaces the optical laws of reflection and refraction are obeyed. These basic and well-known laws can be derived in precise terms from the boundary-value problem of the incidence of a plane wave at any angle to a plane interface between media with differing refractive indices. In the approximation of geometrical optics it is assumed to hold in the locality of a non-planar surface at which a ray is incident. The normal to the surface, the incident, refracted and reflected rays are all coplanar. The surface is considered to be the infinite tangent plane at the point of contact of the ray.

The most basic form of the law of refraction is the bipolar differential form

$$dr = \pm \eta d\rho \qquad \eta = \eta_2/\eta_1 \qquad (1.1)$$

where, as shown in Figure 1.1(a), r and ρ are radial coordinates from the source S and from the point of intersection F of the refracted rays respectively. For well-behaved surfaces this latter point exists in the limit of infinitesimal ray displacements. The ambiguity in sign stems from the manner in which r and ρ are either increasing together or increasing and decreasing respectively. This leads directly to the condition that either a real intersection of the refracted rays occurs and the negative sign is appropriate, or a virtual intersection occurs with the positive sign. Equation 1.1 derives directly from Fermat's principle for the variation of the optical path between the points S and F, for taking the two paths between them we have

$$\eta_1(r + dr) + \eta_2\rho = \eta_1 r + \eta_2(\rho + d\rho)$$

resulting in equation 1.1.

That this relation leads to the more commonly known form of Snell's law can be deduced immediately by dividing equation 1.1 by the arc element ds from

$$ds = (dr^2 + r^2 d\theta^2)^{\frac{1}{2}} = (d\rho^2 + \rho^2 d\phi^2)^{\frac{1}{2}}$$

with the result

$$\eta_1 \sin \theta = \eta_2 \sin \phi \qquad (1.2)$$

In the case of reflection we put $\eta = -1$ and equation 1.1 becomes the even more simple

$$dr = \pm d\rho \qquad (1.3)$$

In this particular case a further simple relationship arises which can be seen from

1

2

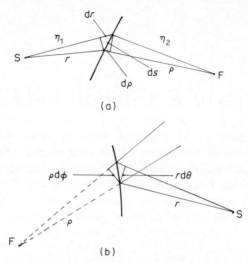

Figure 1.1 (a) The law of refraction $\eta_1 dr = \pm\eta_2 d\rho$
(b) The law of reflection $rd\theta = \pm d\phi$

the congruence of the elemental triangles in Figure 1.1(b), that is

$$rd\theta = \rho d\phi \tag{1.4}$$

From these basic results the totality of geometrical optical design and ray-tracing procedures can be derived.[1]

More importantly, the forms given above can be cascaded to a system of reflecting and refracting surfaces by applying them to successive surfaces and using a multi-polar coordinate system. The boundary conditions applying to the integration of these equations will be the conditions applying to one known ray of the system. This will allow the reduction of the multi-polar coordinate system to a single chosen coordinate system.

We illustrate these properties by applying the fundamental forms to 'designing' conic reflectors. We require in the first instance an axially symmetric surface which will reflect all the rays from a source F into a direction parallel to the axis. Taking the axis of z perpendicular to the axis of symmetry we have (see Figure 1.2) from equation 1.3

$$rd\theta = dz \qquad z = r\sin\theta$$

Thus

$$dr/r = d\theta(1 - \cos\theta)/\sin\theta$$

$$\log r = \log(c\sec^2\theta/2)$$

$$r = 2c/(1 + \cos\theta)$$

a parabola with focal length c.

A fundamental difference occurs between the use of equations 1.3 and 1.4 in this manner. For a reflector which converts all the rays from a source at F_1 into a

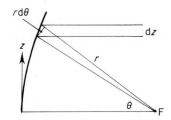

Figure 1.2 Designing the paraboloid from $r\mathrm{d}\theta = \mathrm{d}z$

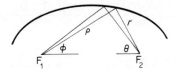

Figure 1.3 Focal property of the ellipse, from $\mathrm{d}r = -\mathrm{d}\rho$

focused system at F_2 as in Figure 1.3, direct integration of equation 1.3 gives, for a real focus

$$r = -\rho + \text{constant} \tag{1.5}$$

or $r + \rho = 2a$, the bipolar equation of the ellipse. This reduces to the polar equation by the application of the (obvious) geometry

$$r \sin \theta = \rho \sin \phi$$

$$r \cos \theta + \rho \cos \phi = F_1 F_2 = 2a\varepsilon$$

ε being the eccentricity.

However, if the second of these is differentiated and equation 1.4 applied, we have the additional properties

$$\frac{\mathrm{d}\theta}{\mathrm{d}\phi} = -\frac{\sin \theta}{\sin \phi}$$

or

$$\frac{\sin \theta}{1 + \cos \theta} \frac{\sin \phi}{1 + \cos \phi} = \text{constant} = A$$

and

$$r = \frac{4Aa\varepsilon}{(1 - A^2) - (1 - A^2)\cos \theta}; \qquad A = \frac{1 - \varepsilon}{1 - \varepsilon}$$

the polar equation of the ellipse, and the relation

$$\frac{\sin \theta}{\sin \phi} = \frac{(1 - \varepsilon \cos \theta)}{(1 - \varepsilon \cos \phi)}$$

Particularly simple combinations based on the reflecting concave hemisphere can be obtained for a variety of focusing requirements. To focus an incoming

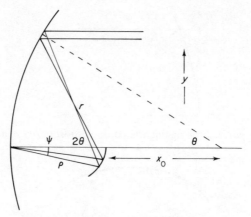

Figure 1.4 Sub-reflector for focusing in a sphere

parallel system of rays at the apex as in Figure 1.4, the three relations available are (for a unit radius hemisphere)

$$\rho + r + \cos \theta = \text{constant} = a \tag{1.6a}$$

from the constancy of ray paths, and

$$x = \cos \theta - r \cos 2\theta \tag{1.6b}$$

$$|y| = r \sin 2\theta - \sin \theta \tag{1.6c}$$

from the geometry of reflection in the sphere with the centre as origin. Equations 1.6b and 1.6c apply to all sub-reflectors contained within the hemisphere.

The constant in equation 1.6a can be evaluated from the condition when $\theta = 0$ to be $a = 3 - 2x_0$. Since

$$\rho \cos \psi = 1 - \cos \theta + r \cos 2\theta$$
$$\rho \sin \psi = r \sin 2\theta - \sin \theta \tag{1.6d}$$

then squaring and eliminating ρ^2 between equations 1.6d and 1.6a gives

$$r = \frac{1}{2} \left[\frac{2 - 2 \cos \theta - (a - \cos \theta)^2}{2 \cos \theta - a - \cos 2\theta} \right] \tag{1.7}$$

and hence x and y are given parametrically from equations 1.6b and 1.6c.

Equations 1.6d and 1.7 can be generalized for a focal point at a distance d from the apex to give

$$r = \frac{1 + (1 - d)^2 - 2(1 - d) \cos \theta - (a - \cos \theta)^2}{2[2 \cos \theta - a - \cos 2\theta(1 - d)]} \tag{1.8}$$

In the same way a telescopic system of rays shown in Figure 1.5 can be obtained from an appropriately shaped sub-reflector. The ray path equation becomes

$$r - x + \cos \theta = \text{constant} \tag{1.9}$$

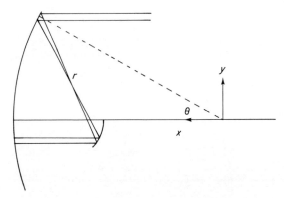

Figure 1.5 Telescopic sub-reflector for the sphere

and from the axis ray at $\theta = 0$ the constant is $2 - 2x_0$. This combined with equations 1.6b and 1.6c gives simply

$$r = a/(1 + \cos 2\theta) \tag{1.10}$$

and the profile parametrically as before.

The required ray property in this case requires that $dy/dx = \cot \theta$, which can be deduced from the parametric equations.

Finally we shall design a sub-reflector profile for the hemisphere which reflects all rays to be perpendicular to the axis. As shown in Figure 1.6, a symmetrically placed sub-reflector would then give the combination the property of retro-reflecting every ray into the symmetrically opposite ray. In this case the ray path equation is

$$r + y + \cos \theta = 2 - x_0 \tag{1.11}$$

giving, together with equations 1.6b and 1.6c

$$x = \frac{1}{2 \cos \theta} - \frac{\cos 2\theta[(4 - 2x_0) \cos \theta - \cos 2\theta - 2]}{2 \cos \theta(1 + \sin 2\theta)}$$

$$y = \frac{\sin \theta}{1 + \sin 2\theta}[(4 - 2x_0) \cos \theta - \cos 2\theta - 2] \tag{1.12}$$

The reflection property here requires that

$$\frac{dy}{dx} = \frac{1 - \tan \theta}{1 + \tan \theta} \tag{1.13}$$

and, although this is difficult to derive from the differentiation of equation 1.12, it can be integrated to derive equation 1.12 itself. As shown, the curve obeys the ray requirement geometrically, even for rays which cannot physically reach it through being intercepted beforehand by the main hemispherical reflector.

In all these examples we have taken the position of the sub-reflector to be x_0 from the centre of the hemisphere. In the case of the hemisphere reflecting a beam

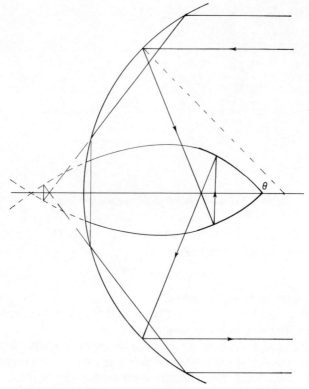

Figure 1.6 Sub-reflector for retro-reflection in the sphere

of parallel rays, the well-known caustic (Figure 1 of Appendix I) occurs. This will be dealt with more fully subsequently (see section 2.5.3). When the sub-reflector is positioned between the caustic and the reflector, that is when $x_0 > 0.5$, two rays can pass through each point (the two tangents to the upper half of the caustic; a third tangent exists but we here consider the symmetrical half of the problem). In this region no single reflector can redirect both rays since they arrive at different angles. However, in this region the above curves all become double-valued, giving a double-branched curve, one branch for each of the ray pencils incident in that region. The development of the double-branched curve as the starting value moves through the cusp of the caustic is shown in Figure 1.7. The two rays that pass through P are reflected from different branches of the curve where it is double-valued, to focus correctly at the apex. Hence no reflection occurs at Q.

Turning now to refraction, we can produce the same exact focusing effect with a single surface, the analogue of the parabola, by integrating equation 1.1 directly. This gives

$$r = -\eta\rho + \text{constant} \tag{1.14}$$

The constant is obtained as usual by the condition for a limiting path, and thus the constant equals $f_1 + \eta f_2$ (Figure 1.8). In general equation 1.13 is that of a

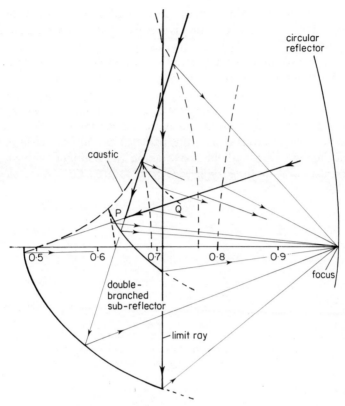

Figure 1.7 The development of the double-branched sub-reflector. The two rays through
P reflect off different branches of the curve to focus at the apex

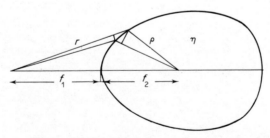

Figure 1.8 The Maxwell oval, from $dr = -\eta d\rho$

Cartesian or Maxwellian oval[2] having different forms depending on the chosen
positions of the foci and the value and *sign* of η.

We have taken the refraction to be from a source in a lower refractive medium.
This path is reversed for a source in the higher refractive medium, but the roles of r
and ρ can then be reversed also. So for $\eta < 1$ we use $1/\eta > 1$ and replace
$dr = \pm\eta d\rho$ with

$$\rho = \pm\eta dr$$

8

The situation reversal that this implies can be used in all resultant formulae, except in situations where the refraction is singular, that is the condition of total internal reflection.

Snell's own construction for refraction can be extended to cover multiple surfaces and total internal reflection. The basic construction is as shown in Figure 1.9. The circles have radii proportional to the ratio of the refractive indices and the incident and refractive rays are derived by the construction shown. Applying the sine law to the triangle OPQ shows the agreement with Snell's law of refraction. For a multilayer system, the refracted ray in one medium is the incident ray for the next, and the construction can continue with piecewise straight segments and abutting triangles OP_1Q_1, OQ_1P_2, OP_2Q_2 etc. as shown in Figure 1.9.

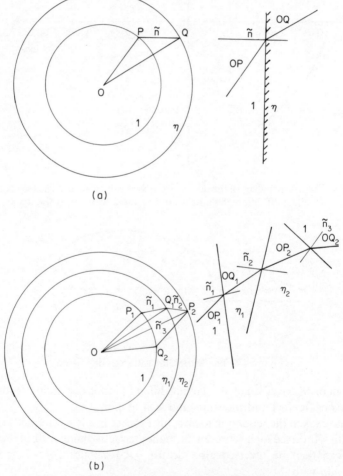

(a)

(b)

Figure 1.9 (a) Snell's construction for refraction
(b) Snell's construction for several surfaces

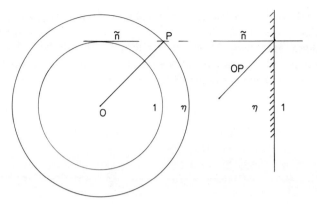

Figure 1.10 Snell's construction for total internal reflection

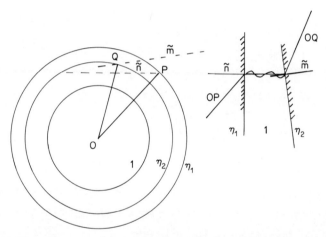

Figure 1.11 Tunnelling rays in total internal reflection. ñ intersects the unit circle in complex conjugate points, and m̃ passes through this intersection and intersects the circle η_2 at the real point Q

If the ray is incident from the side with the higher refractive index, the condition for total internal reflection, $\sin\theta = 1/\eta$, is derived immediately (Figure 1.10).

We take now the case where two media with refractive indices η_1 and η_2 are separated by a small air gap as shown in Figure 1.11. The ray is incident at an angle beyond the critical value and is totally internally reflected. That is, the normal ñ through P does not intersect the unit circle in real points. It does, however, do so at two conjugate imaginary points, given by the intersection of the line $y = \eta_1 \sin\theta$ with the circle

$$x^2 + y^2 = 1$$

that is, at points $x = \pm i(\eta_1^2 \sin^2\theta - 1)$, $y = \eta_1 \sin\theta$.

If the normal m̃ to the second surface passing through one of these points has a *real* intersection with the circle $x^2 + y^2 = \eta_2^2$, then a real ray will exist in the second medium as shown. However, its amplitude will be reduced by the exponential decay of evanescent fields along the path PQ, and this too can be obtained from the geometrical construction given.

2 The Zero-distance Phase Front

2.1 DEFINITION AND GENERAL DERIVATION

Rays in an homogeneous isotropic medium, which originate from a point source, form a normal congruence. That is, if equal optical paths are measured along each ray from the source, the surface constructed by the end points will be normal to all of the rays in the congruence. It follows that a similar result will be obtained if the distances along each ray are measured from any one of the normal surfaces themselves. These surfaces are the phase fronts of the wave system, or orthotomics, for which the rays are the geometrical optics approximation. A fundamental theorem, the theorem of Malus and Dupin,[3] states that a normal congruence will remain a normal congruence after any number of reflections or refractions. The phase front surfaces, however, will vary considerably at each intersection of the ray system with a reflecting or refracting surface. In an isotropic non-homogeneous medium the phase front is a continuously deformable surface normal to the now curving ray system.

We define a single special phase front relating to a specific combination of source and retracting surface.[4] This is constructed by the process illustrated in Figure 2.1. A general ray from the source S is incident upon the surface g, between two homogeneous media with refractive indices η_1, containing the source S, and η_2, at the point M. At M it undergoes refraction in accordance with Snell's law

$$\eta_1 \sin \theta = \eta_2 \sin \phi$$

where θ and ϕ are the incident and refracted ray angles with the normal to the surface at M. If the refracted ray is continued in its *reverse* direction to a point P, where

$$\eta_2 \text{MP} = -\eta_1 \text{SM}$$

the minus sign indicating the reversal of direction, the points P will all lie on a surface h, as M moves over the surface g. This surface h is defined to be the 'phase front of zero distance'. As a consequence of the theorem of Malus and Dupin, all the rays MP are normal to h, and the optical distance from the source S to P is zero. It is thus the virtual-*source* phase front for the image space to the right of g.

In most cases to be considered, the source will be in free space and the refracting medium uniform with refractive index η. Then with

$$\eta_1 = 1 \qquad \eta_2 = \eta$$

the surface h will be defined by

$$\eta \text{MP} + \text{SM} = 0 \tag{2.1}$$

11

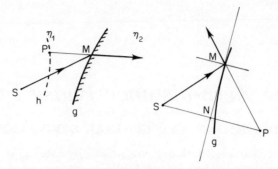

Figure 2.1 Derivation of the zero-distance phase front. The ray refracted at M is produced to P by a distance $-(\eta_2/\eta_1)$SM. The locus of P is the curve h, the zero-distance phase front. In the case of reflection $\eta_2/\eta_1 = -1$ and P lies on the opposite side to S and is the reflection of S in the tangent at M. The locus of N is the (first) pedal curve of g

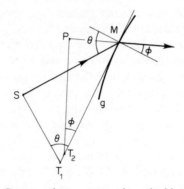

Figure 2.2 Construction to prove the coincidence of T_1 and T_2

For the case of a reflector we put $\eta = -1$ in the last instance. P then lies on the *opposite* side of the reflector from S, and MP = SM. P is thus the reflected point of S in the tangent at M.

If now, as shown in Figure 2.2, we draw the tangent to g at M, then the perpendicular to SM at S will meet it at T_1 and the perpendicular to PM at P will meet it at T_2. The angle ST_1M then equals θ and the angle PT_2M equals ϕ as shown. Therefore by the construction for P

$$\frac{PM}{T_2M} = \sin\phi \qquad \frac{SM}{T_1M} = \sin\theta$$

But SM/PM $= \eta = \sin\theta/\sin\phi$; therefore $T_1M = T_2M$, and so T_1 and T_2 coincide at T, the result shown in Figure 2.3.

This construction makes possible a general parametric description of the zero-distance phase front without recourse to point-by-point ray-tracing methods.

Taking the source at the origin of a plane coordinate system in which the refracting profile g is given parametrically by

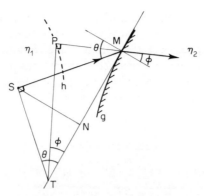

Figure 2.3 Geometrical construction of the locus of P. P is the intersection of the circle with diameter MT and the circle with centre M and radius (η_2/η_1)SM

$$x = f(t) \qquad y = g(t)$$

the tangent at $M(x, y)$ has equation

$$y - g(t) = \frac{g'}{f'}(x - f(t)) \tag{2.2}$$

where primes refer to differentiation with respect to the parameter t (and in all following sections).

T is the intersection of this line with the perpendicular from the origin to SM, or

$$g(t)y = -f(t)x$$

Thus T has coordinates

$$X = \frac{fgg' - g^2f'}{gg' + ff'} \qquad Y = \frac{fgf' - f^2g'}{gg' + ff'} \tag{2.3}$$

P is now *one* of the intersections of the circle with centre M and radius PM = SM/η, and the circle with MT as diameter, that is the circles

$$(x - f)^2 + (y - g)^2 = (f^2 + g^2)/\eta^2$$

$$[x - (X + f)/2]^2 + [y - (Y + g)/2]^2 = \tfrac{1}{4}[(X - f)^2 + (Y - g)^2]$$

with the result, putting $N = gg' + ff'$

$$\eta^2(x - f)(f'^2 + g'^2) + Nf' = \pm g'[\eta^2(f^2 + g^2)(f'^2 + g'^2) - N^2]^{\frac{1}{2}}$$
$$\eta^2(y - g)(f'^2 + g'^2) + Ng' = \mp f'[\eta^2(f^2 + g^2)(f'^2 + g'^2) - N^2]^{\frac{1}{2}} \tag{2.4}$$

The ambiguity in sign shows the existence of the second intersection of the two circles, and care has to be taken to obtain that intersection which agrees with the physical situation.

In general these equations are too complicated for the elimination of the

14

parameter t, to enable the derivation of a functional relation between x and y alone. They are, however, quite suitable for computational methods. The result shows that a purely geometrical description of the zero-distance phase front is possible from a given geometrical surface without considering individual rays.

A simple illustration of this result is the zero-distance phase front of a plane interface. For this the plane is parametrized by

$$f(t) = c \qquad g(t) = t$$

where c is a constant. Equations 2.4 then give

$$\eta^2(x - c) = \pm[(c^2 + t^2)\eta^2 - t^2]^{\frac{1}{2}}$$
$$\eta^2(y - t) = -t$$

which on eliminating t is

$$(x - c)^2 - \frac{y^2}{\eta^2 - 1} = \frac{c^2}{\eta^2} \qquad \text{(fig. 2.4)} \qquad (2.5)$$

This is a hyperbola if $\eta > 1$ and the refraction is from a source in free space, or an ellipse if $\eta < 1$ and the source is within the refractive medium.[5] In either case the source is at the focus of the conic and the plane interface passes through the centre.

2.2 REFRACTION IN A CIRCULAR INTERFACE

For the all-important case of refraction in a circular interface we utilize the geometry of Figure 2.3 but with g a circle of radius r, whose centre O is a distance k from the source S (Figure 2.5). Then equating the trigonometric identities for the triangles SPO, SMO and MPO, and with $SM = \eta PM$, we obtain respectively

$$SP^2 = \rho^2 = SM^2[1 + \eta^2 - 2\eta \cos(\theta - \phi)]/\eta^2$$
$$k^2 = SM^2 + r^2 + 2SMr \cos\theta$$

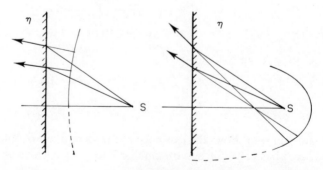

Figure 2.4 Construction of the zero-distance phase fronts for refraction in a plane interface

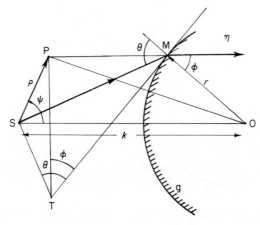

Figure 2.5 Construction of the zero-distance phase front for refraction in a circular interface

$$\mathrm{PO}^2 = k^2 + \rho^2 - 2k\rho \cos \psi$$
$$= r^2 + \mathrm{SM}^2/\eta^2 + 2\mathrm{SM}r \cos \phi/\eta$$

It can readily be shown that $\rho = \pm \mathrm{SM}(\cos \phi - \cos \theta/\eta)$ (by squaring and comparing with ρ^2 above), and hence SM and θ can be eliminated to give the polar (ρ, ψ) equation of P with respect to S; that is

$$\eta^2\rho^2 - 2\eta\rho(r + \eta k \cos \psi) + (\eta^2 - 1)(k^2 - r^2) = 0 \qquad (2.6)$$

The ambiguity in the sign of ρr has been resolved by the consideration of the case where g is a *reflecting* surface as shown in section 2.3.

In performing this analysis we used only the fact that the normal to the surface at M goes through the centre O. Since the refraction of the ray at M is only a *local* effect, the same geometry results if g were taken to be the circle of curvature with double contact at M. This leaves the position of the tangent MT to the surface and to the circle of curvature, unaltered. Hence we can replace r and k in equation 2.6 by R and p (Figure 2.6), where R is the radius of curvature and p the distance from S of the centre of curvature C at any point M on the now general surface g.

Thus the zero-distance phase front with respect to axes at S is given by (ρ, Ψ), where

$$\eta^2\rho^2 - 2\eta\rho(R + \eta\rho \cos (\Psi - \varepsilon)) + (\eta^2 - 1)(p^2 - R^2) = 0 \qquad (2.7)$$

From the standard literature, if the surface g has polar coordinates (s, μ) with respect to S, then

$$R = \frac{(s^2 + s'^2)^{\frac{3}{2}}}{s^2 + 2s'^2 - ss''}; \qquad s' = \frac{\mathrm{d}s}{\mathrm{d}\mu} \qquad s'' = \frac{\mathrm{d}^2s}{\mathrm{d}\mu^2}$$

and the coordinates of C are

16

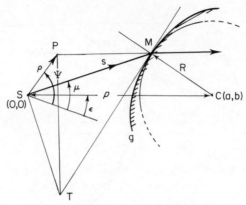

Figure 2.6 Construction for a general surface g. The circle with centre C is the circle of curvature of the point M

$$a = s \cos \mu - \frac{s^2 + s'^2 d(s \sin \mu)}{s^2 + 2s'^2 - ss''}$$

$$b = s \sin \mu + \frac{s^2 + s'^2 d(s \cos \mu)}{s^2 + 2s'^2 - ss''}$$

$$\varepsilon = \tan^{-1}(b/a)$$

We can recover the condition for a plane interface by a method which will have other applications. Equation 2.6 can be rewritten as

$$\eta^2 \rho^2 - 2\eta\rho[(r-k) + k(1 + \eta \cos \psi)] - (\eta^2 - 1)[2k(r-k) + (r-k)^2] = 0$$

(2.8)

Then r and k can become infinitely large while the difference $(r - k)$ remains finite and equal to $-C$. In the limit

$$\rho = \frac{(\eta^2 - 1)C}{\eta(1 + \eta \cos \psi)}$$

which is the polar equation of equation 2.5.

We now consider the general conditions of the curve of equation 2.6 for different positions of the source, that is for varying values of k and r, and for refraction both from the source interior to and exterior to the refracting circle.

In general the curve is double-branched, each branch being the result of the refraction in the convex or concave parts of the circle respectively (Figure 2.7). The source S is the focal point for the former and there is a second pole V relating to the concave refraction. The distance SV is given by

$$SV = k(\eta^2 - 1)/\eta^2$$

(2.9)

The bi-polar equation to the curve in terms of radial coordinates r_1 from S and r_2 from V is

$$r_1 - \eta\{r_2 \pm r(\eta^2 - 1)/\eta^2\} = 0$$

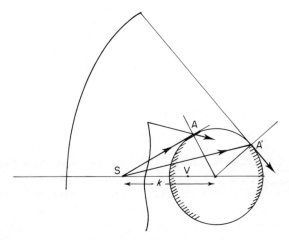

Figure 2.7 Zero-distance phase fronts for the convex and concave parts of a circular interface

or

$$\eta r_2 - r_1 = \pm \eta r/k \qquad (2.10)$$

Hence when $r = k$, that is the source is on the circle itself, the zero-distance phase front is composed of the isolated point S and the limaçon of Pascal[6,7]

$$\rho = 2r(1 - \eta \cos \psi)/\eta \qquad (2.11)$$

For the refraction from a source in free space η is greater than unity. Then with the source within the circle, $0 < k < r$, the curve of equation 2.6 is closed and entirely within the circle. It intersects the axis at points

$$(k - r)(1 - 1/\eta) \qquad (k + r)(1 - 1/\eta)$$

With the source exterior to the circle the curve has two parts, one interior to the other, corresponding to the parts of the circle for which the surface is convex or concave towards S.

When $k = \eta r$ the interior curve degenerates to the point V.

For refraction from a source *inside* a circular refracting medium, η is less than unity and rays from the source are limited by the condition of total internal reflection to within an angle W such that

$$\sin W < r/\eta k$$

Thus if $k < r/\eta$ this condition will not occur and all rays will leave the circle.

If $k = r/\eta$ all the rays from S are concurrent at V which is at a distance ηr from the centre of the circle. This is the condition for the aplanatic points of the sphere.[2]

Thus the form of the zero-distance phase front for refraction in a circular interface is seen to be intimately involved with the form taken by different parametric values of the limaçon of Pascal.

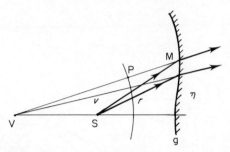

Figure 2.8 Surface required for a circular zero-distance phase front

We conclude by the determination of that surface profile that would produce a *circular* zero-distance phase front, by refraction from a source S in free space. To do this we require the surface that would give a *virtual* focus for all the rays at V, which is then the centre of the phase front. If then, as in Figure 2.8, the law of refraction at M is put into its differential form (equation 1.1)

$$dr = \eta dv$$

where r and v are bipolar coordinates from S and V respectively, direct integration gives

$$r = \eta v + D \qquad D = f - \eta(a + f) \tag{2.12}$$

where the constant of integration is obtained by the condition for the central ray from S to V. Equation 2.12 is the bipolar equation of the limaçon. Hence the situation is that refraction in a circle produces a zero-distance phase front which is a limaçon and refraction in a limaçon produces a phase front which is a circle. We have already met a similar situation in the limit of the plane surface refraction. Equation 2.5 showed this to be either elliptical or hyperbolic depending upon whether the source was inside the medium or in free space. Now, as shown in Figure 2.9, if a source is embedded within a medium, then the ellipse is the single refracting surface that will produce parallel rays that is a plane phase front. Likewise a source in free space will be collimated to parallel rays by the single refraction in a hyperbolic surface. Hence a *plane* refractor gives a hyperbolic or elliptical zero-distance phase front, and the hyperbolic or elliptical refractor gives a plane phase front.

2.3 REFLECTION IN A CIRCLE

With the exception of the singular, and obvious, case of the plane surface, reflection can be dealt with by putting $\eta = -1$ in all of the foregoing analysis. Thus for the circular reflector, from equation 2.6, we get immediately

$$\rho = 2(k \cos \psi - r) \tag{2.13}$$

which again is the equation of a limaçon, also taking various forms as the ratio of k and r varies.

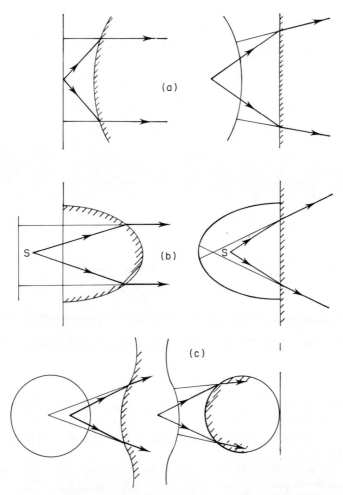

Figure 2.9 (a) Refraction in a hyperbola produces a plane phase front and refraction in a plane a hyperbolic phase front
        (b) With the source in the interior the same effect occurs with elliptical surfaces
        (c) The symmetrical situation for the circle and limaçon, the inversion of (b) in the local source S
In (a) and (b) the two conditions shown are reflections, which are inversions with centre at infinity

If, as in Figure 2.10, O is a fixed point and OP a line intersecting a curve C at a point Q, then the locus of points P_1 and P_2 such that

$$P_1Q = P_2Q = \text{constant}$$

are each a *conchoid* of C with respect to O. Equation 2.13 expresses the result that the zero-distance phase front of a source in a reflecting circle is the conchoid of one circle radius k with respect to a point on its circumference. It is also the *epitrochoid* generated by the fixed point on one circle, rolling without slipping

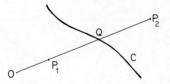

Figure 2.10 Construction of the conchoid of the curve C

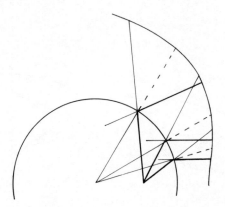

Figure 2.11 The epitrochoid as the conchoid of the circle

upon another equal fixed circle (Figure 2.11). When $r = k$ the source is on the circumference and the curve becomes a *cardioid*. When $r = k/2$ it is a *trisectrix* (Figure 2.12). The nature of all these curves is described in the table of curves given in Appendix I.

If $r > k$ the source is inside the circle and the limaçon is a single loop. If $r < k$ the limaçon forms two loops which are the result of reflection in the respective parts of the circle convex or concave towards the source.

2.4 REFLECTION IN A GENERAL CURVE

For a reflector of any shape whose profile is the curve $f(x, y) = 0$, we can obtain the zero-distance phase front for a source at any location (d, h) by direct means. We erect a cylindrical reflector with generators parallel to the z axis from the cross-section (Figure 2.13). Then a ray from the source S to the point (a, b, c) will, after reflection, intersect the plane $z = 0$ in a point P, on the zero-distance phase front, and independently of c. The direction of the ray can be derived directly from the vector form of the law of reflection[8]

$$\hat{S}_r = \hat{S}_i - 2\hat{n}(\hat{S}_i . \hat{n}) \tag{2.14}$$

with unit vectors shown in Figure 2.13. Since the normal $\hat{n}$ at (a, b, c) is given by

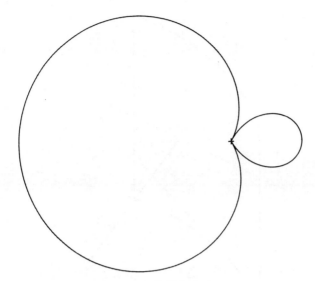

Figure 2.12 Trisectrix

∇f, then P has coordinates

$$x = d + 2 \frac{\partial f}{\partial x} A/B$$

$$y = h + 2 \frac{\partial f}{\partial y} A/B$$

$$A = \left[(a - d) \frac{\partial f}{\partial x} + (b - h) \frac{\partial f}{\partial y} \right]$$

$$B = \left[\left(\frac{\partial f}{\partial x} \right)^2 + \left(\frac{\partial f}{\partial y} \right)^2 \right] \qquad (2.15)$$

where all the differentials are evaluated at the point (a, b). Since (a, b) also satisfies $f(a, b) = 0$, it is possible to eliminate them from equations 2.15 and derive an implicit relation for the zero-distance phase front.

As an example we shall derive by this method the zero-distance phase front of a parabola with the extension required to deal with the possibility of having the source at an infinite distance. That is the zero-distance phase front for plane wave sources. We know, of course, that for the parabola with its focusing property, this will be a circle, strictly of infinite radius, centred on the focal point.

Using equations 2.15 for the parabola with focal length a

$$f(x, y) = y^2 - 4a^2 + 4ax = 0$$

and a source on the straight line $x = 0$, $y = h$, the curve of intersection with the plane $z = 0$ is

$$F(x, y, h) = [y(x - 2a) - 2ha]^2 - x^2 h^2 - (x - 2a)^2 = 0 \qquad (2.16)$$

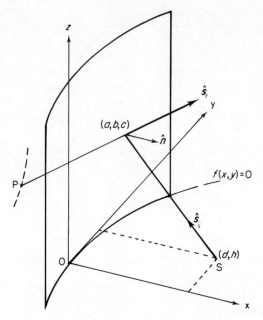

Figure 2.13 Zero-distance phase front of a general reflecting surface

This, incidentally, is the true curve of the phase front for an offset feed in a paraboloid and its digression from a straight line exhibits the cause of coma aberration in optical systems.[9]

For an incident *plane* wave we require the envelope of this equation with varying h, that is we need to eliminate h between equation 2.16 and the equation $\partial F/\partial h = 0$. This results in $h = 2ay/(2a + x)$ which, substituted back into equation 2.16, gives the circle

$$x^2 + y^2 = 4a^2$$

This radius is *not*, however, infinite, since we have not derived the zero-distance phase front but the phase front due to zero phase along the line $y = h$. For circular phase fronts, zero distance becomes a matter of defining the radius only.

The method illustrates that a phase front in general, and a zero-distance phase front in particular, can be obtained as the *envelope* of phase fronts from a moving point source. This is thus the process required to obtain the zero-distance phase fronts in those cases where the source is a *caustic*, or another proposed phase front. This envelope procedure, however, does *not* apply to the reflecting or refracting surfaces themselves.

We note also that equations 2.6 take on a particularly simple form for a reflector when $\eta = -1$. Then if the reflector is given parametrically by

$$x = f(t) \qquad y = g(t)$$

the zero-distance phase front is

$$x = 2(fg'^2 - gf'g')/(f'^2 + g'^2)$$
$$y = 2(gf'^2 - ff'g')/(f'^2 + g'^2)$$

(2.17)

These coordinates are double in value to those of the point N in Figure 2.1, which is the foot of the perpendicular from S onto the tangent at M, as would be expected in the case of a reflector. The locus of N is defined to be the (first positive) *pedal* curve of g with respect to S.

Hence the zero-distance phase front of a reflector can be derived directly from the pedal curve of the reflecting profile. With a source at the point (d, h) and the parametric form of the reflector profile this construction, and equation 2.6, gives the zero-distance phase front

$$x = \frac{df'^2 + 2fg'^2 + 2(h - g)f'g' - dg'^2}{f'^2 + g'^2}$$
$$y = \frac{hg'^2 + 2gf'^2 + 2(d - f)f'g' - hf'^2}{f'^2 + g'^2}$$

(2.18)

the primes, as always, referring to differentiation with respect to the parameter t.

A table of pedal curves is included in the geometrical constructions given in Appendix II.

2.5 GEOMETRICAL CONSTRUCTIONS

We shall require, for future operations with zero-distance phase fronts, some geometrical results which are standard but are included for completeness. Most of the details, and the tables in Appendices I and II, were derived from the references listed in Reference (6), mainly from the first of those given.

2.5.1 Parallel surfaces

As already stated, the theorem of Malus and Dupin states that the propagating wave, or phase front in a uniform medium, is everywhere normal to the congruence of rays. Since the zero-distance phase front is a source of such a congruence, all ensuing phase fronts are 'parallel' in the geometrical sense. That is, points on each surface are equidistant where the distance is measured along the common normal between points on each of the surfaces. If P is on a phase front whose profile is given parametrically by $f(t), g(t)$, then $Q(x, y)$ will be on a parallel phase front at a distance l (Figure 2.14) given by

$$x = f(t) \pm lg'(f'^2 + g'^2)^{-\frac{1}{2}}$$
$$y = g(t) + lf'(f'^2 + g'^2)^{-\frac{1}{2}}$$

(2.19)

We can see by inspection that the only form-invariant phase fronts are the sphere and the plane. The same relation applies when considering Huygens' construction for propagating wavefronts in a uniform medium.

24

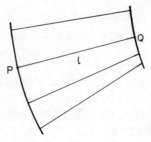

Figure 2.14 Construction of the parallel of a curve PQ = *l*

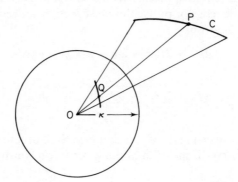

Figure 2.15 Construction of the inverse of a curve OP.OQ = κ^2

2.5.2 Inversion

The radius vector through the origin O, which is the *centre of inversion*, intersects a curve C at a point P (Figure 2.15). The point Q on OP such that

$$OP.OQ = \kappa^2$$

is the inverse point of P with radius of inversion κ. The locus of Q as P moves on C is the inverse curve of C.

It is also important to define the negative inverse

$$OP.OQ = -\kappa^2$$

in which case P and Q are taken to be on opposite sides of O.

If O is the pole of a polar curve $r = f(\theta)$, the inverse with respect of the pole is given by

$$Rr = \kappa^2$$

or

$$R = \kappa^2/f(\theta)$$

θ remaining invariant in this process.

If O is the point (a, b) and the curve C is given parametrically by $x = f(t)$, $y = g(t)$, the equation of the inverse curve $Q(x, y)$ is

$$x = a + \frac{\kappa^2[f(t) - a]}{\{[f(t) - a]^2 + [g(t) - b]^2\}}$$

$$y = b + \frac{\kappa^2[g(t) - b]}{\{[f(t) - a]^2 + [g(t) - b]^2\}}$$

(2.20)

Under the inversion the angles between curves are unaltered and the mapping is *conformal*. Circles not passing through the centre of inversion invert into circles, and a circle passing through the centre inverts into a straight-line tangent to the circle at the centre of inversion. Similarly asymptotes invert into tangents of the inverse curve at the centre of inversion. Some curves such as the spirals are self-inverse and are termed *anallagmatic*. The inverse of an inverse in the same centre and with the same radius of inversion returns the original curve.

For inversion when the centre of inversion is at infinity, a requirement we shall later use, we adopt the convention that inversion is a reflection in *any* local plane whose normal is in the direction of the point at infinity specified as the centre of inversion.

The inversion transformation in three-dimensional space was pointed out by Maxwell[10] to be the only conformal mapping which transforms three-dimensional space into itself. The same can be shown to be true of four-dimensional space. We ignore, of course, the 'obvious' transformations, such as translation, rotation and reflection. Bateman[11] also shows that it is the general form for transformations in four dimensions leaving the wave equation and the eikonal equation invariant, and Cunningham[12] shows the same role for it in transformations of the electromagnetic field. Thus its applications to geometrical optics could be anticipated in these studies, as we will indicate subsequently.

2.5.3 Caustics, evolutes and involutes

The caustic of a given curve C is the envelope of the rays emitted from the source S after reflection in the curve, the *catacaustic* or, after refraction, the *diacaustic*. From Figure 2.16, P is the reflection of S in the tangent to the surface at M for the case of a reflection, and MQ = MP. Since MQ is the reflected ray, it is normal to the locus of P and thus Q is the centre of curvature of the zero-distance phase front. The locus of the centre of curvature of a curve is the evolute of the curve. Thus Q describes the *caustic by reflection* in g, which at the same time is the evolute of the zero-distance phase front.

The evolute of a curve with parametric definition $x = f(t)$, $y = g(t)$ is

$$x = f(t) - (f'^2 + g'^2)g'/(f'g'' - f''g')$$

$$y = g(t) + (f'^2 + g'^2)f'/(f'g'' - f''g')$$

(2.21)

A further derived curve from a given curve C is obtained from the motion of a point P fixed to the line L which rolls without slipping along the curve C. The same curve is obtained as if P were fixed to a taut inextensible string attached to a point on C and winding or unwinding over the curve, as shown in Figure 2.17.

26

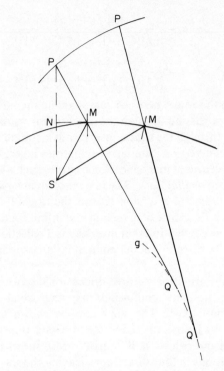

Figure 2.16 Construction of the catacaustic of a curve. Q is the centre of curvature of the curve at P

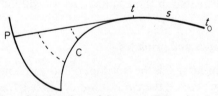

Figure 2.17 Construction of the involute of the curve C. Ptt_0 is a taut string attached to C at t_0 and wound about it to t

The part of the string forming the tangent to the curve replaces the line L of the first description. The locus of P is then the *involute* of the curve C. Different positions of the point P or different positions of the point of fixing of the string to the curve give involutes, every involute described in this manner being a parallel of every other involute. This indicates the connection to be found between involute curves and phase fronts.

If the fixed curve C has parametric definition $x = f(t)$, $y = g(t)$, and the tangent from P meets the curve at a point that has parameter t, then the equation of the involute is

$$X = f(t) - sf'/(f'^2 + g'^2)^{\frac{1}{2}}$$
$$Y = g(t) - sg'/(f'^2 + g'^2)^{\frac{1}{2}}$$

(2.22)

where s is the arc length of that part of the string lying along the curve from the fixed point to the point of tangency t. That is

$$s = \int_{t_0}^{t} (f'^2 + g'^2)^{\frac{1}{2}} dt \qquad (2.23)$$

If a curve A is the involute of a curve B, then B is the evolute of A, and vice versa. Since the caustic is the evolute of the zero-distance phase front, the zero-distance phase front is the involute of the caustic. This gives a further method for its description, since caustics can be obtained as the envelope of reflected or refracted rays.

The description of the involute to a caustic has to take into consideration the double-valuedness of the ray pattern on the illuminated side of the caustic. The commonest observed caustic is that of the plane wave incident upon a circular reflector (see Figure 1.7). Since at each point on the illuminated side two tangents can be drawn to the caustic, each being a ray path, then any involute or phase front, which is nominally perpendicular to each ray, will likewise have two values at that point. The method of drawing the involute shown in Figure 2.17 will only take into account those rays which after reflection are single-valued, and for any practical application the aperture has to be limited to stop off the second pencil of rays. The point of attachment of the unwinding string then decides the nature of the local region of the involute.

2.5.4 Examples

To illustrate these points and the application of the table in Appendix II we can derive the phase fronts for the illumination by a parallel beam of rays of some interesting curved reflectors.

First we shall consider a reflector with a profile given by the curve $y = \ln x$. The process for deriving the caustic is to obtain the envelope of the reflected rays. From Figure 2.18 the following relations apply

$$\tan \theta = \frac{dy}{dx} = \frac{1}{x}$$

The equation of the reflected ray

$$Y - y = \tan 2\theta (X - x)$$

then becomes

$$F(\theta) = Y - \log \cot \theta - \tan 2\theta (X - \cot \theta) = 0$$

and for the envelope

$$\frac{\partial F}{\partial \theta} = 1 - X \sin 2\theta = 0$$

Thus

$$\frac{dY}{dX} = \tan 2\theta = 1/[X^2 - 1]^{\frac{1}{2}}$$

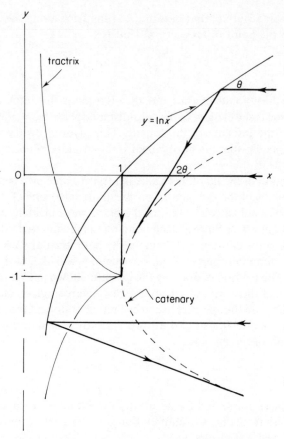

Figure 2.18 Caustic and involute for the curve $y = \ln x$

so that

$$X = \cosh(Y + 1)$$

the constant being obtained from the reflection of the ray $y = 0$ into the ray $x = 1$.

Thus the caustic is the catenary and the phase front the involute of the catenary which is the tractrix. This is double-branched, the upper half showing a 'real' phase front, that is after the rays have been tangential to the catenary, and the lower half a 'virtual' phase front, occurring as it does before the ray has reached the caustic.

In Figure 2.19 the catacaustic of the cycloidal arch with parameter a is shown to be the two equal cycloidal arches as detailed in the tables in Appendix II. The involute of a cycloidal arch is an equal cycloidal arch, and hence the phase front is the latter arch as shown.

In the case of the logarithmic spiral both the catacaustic and its involute, the phase front, are equal logarithmic spirals, as illustrated in Figure 2.20.

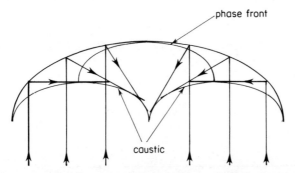

Figure 2.19 Caustic and involute of the cycloidal arch. Both caustic and involute are cycloidal arches with half parameter of the reflecting arch

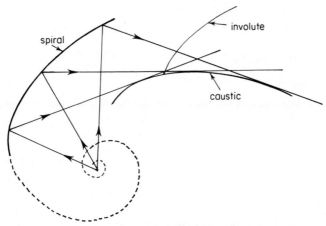

Figure 2.20 Caustic and involute of the logarithmic spiral. All three curves are equal spirals

2.6 CAUSTIC APPROXIMATIONS

2.6.1 Approximation by an equivalent point source

The properties of the evolute and involute as applied to caustics and zero-distance phase fronts give rise to a method of approximation of great value in tracing fronts through systems with several surfaces. The analysis to date has been concerned with the full range of rays emitted by a point source. In practical instances only a specific angular range of rays will need to be considered, and in the paraxial case only a very small angle about the axis of the system. Hence only a small part of the curves concerned, phase fronts or caustics, will in general need to be used, and most often in the region about the axis of symmetry. Thus it will be possible to replace phase fronts over this region by simpler curves, curves that can be used to continue the process through to the next surface. Similarly, caustics can be replaced by simpler curves or even in some cases by a point source. In every case the approximation made by such replacements exactly specifies the

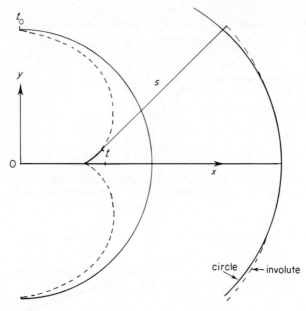

Figure 2.21 Approximation of the phase front, the involute of the caustic of a circular mirror by a circle

amount of aberration *being introduced by the approximation*. The divergence of phase fronts from true circles (or planes if focused at infinity) or caustics from true points exactly specifies the amount of aberration inherent in the optical system.

The method for approximating a caustic to an equivalent point is, in fact, to derive the involute, that is the phase front, and approximate that to a circle. The centre of the circle will then be the best equivalent point for the caustic. Of course there are numerous ways to approximate the curve to the circle—a best mean-square fit, for example. In the examples chosen here we simply fit the circle to three arbitrary selected points on the phase front.

We take first the circular reflector illuminated by a parallel beam of rays (Figure 2.21). The caustic is the well-known cusp of the nephroid with parametric equations (Figure 1.1.1 Appendix I)

$$x = a(3 \cos t - \cos 3t) \equiv f(t)$$
$$y = a(3 \sin t - \sin 3t) \equiv g(t)$$

(2.24)

Without loss of generality a can be taken to be unity; the radius of the circle is then 4. The cusp corresponds to $t = 0$ and the extrema on the y axis to $t = \pm \pi/2$, $y = 4$.

We then have from equation 2.23

$$s = 6(\cos t_0 - \cos t)$$

where t_0 is the starting point for the unwinding string forming the involute. The parametric equation of the involute, that is of the phase front, is then

$$X = 3 \cos t - \cos 3t - 3(\sin 3t - \sin t)(\cos t_0 - \cos t)/\sin t$$
$$Y = 3 \sin t - \sin 3t - 3(\cos t - \cos 3t)(\cos t_0 - \cos t)/\sin t \qquad (2.25)$$

Care has to be taken with the signs of the final terms on the right-hand side owing to the ambiguity in the square-root of the denominators of equations 2.22. The value of t_0 determines the range around the cusp for which the approximation is being sought. If we take the entire (half) nephroid, then $t_0 = \pi/2$. Three points on the curve of equation 2.25 can then be taken to be $t = 0$ and $\pm \pi/4$. Inserting these values we obtain the following three points on the true phase front

$$(8, 0) \qquad (2\sqrt{2}, 4\sqrt{2}) \qquad (2\sqrt{2}, -4\sqrt{2})$$

The centre of the circle through these points is on the x axis at the point

$$x = \frac{3\sqrt{2}}{2\sqrt{2} - 1} \simeq 2.32$$

which is not the half radius usually chosen as the source point for a circular reflector.

The difference between the exact (computable) curve of equation 2.25 and the circle *with this centre* expresses the spherical aberration of the reflector. Different circles are obtained if the range of the comparison is reduced by taking values of t closer to the central value $t = 0$.

We use the same method to determine the best position for a source point in a parabola to give a parallel beam of rays at some angle θ to the axis.[13,14] The caustic is Tschirnhausen's cubic

$$f(t) = -bt(t^2 - 3)/9$$
$$g(t) = bt^2/3$$

where $b = 9a \sin \theta$ and a is the focal length.

This curve has been referred to oblique axes which are the symmetry axes of the cubic (Figure 2.22). The illuminated portion of the paraboloid creates a region of this caustic around the parametric value $t = 0$. The value of t_0 depends on the geometry of the paraboloid; for a full focal plane paraboloid this is $t_0 = \pm 1$. For a paraboloid subtending a total angle 2Θ at the focus, t_0 is approximately $\pm \tan(\Theta/2)$.

The procedure given above leads to the parametric equation to the involute

$$X = \frac{4bt^3 + bt_0(3 + t_0^2)(1 - t^2)}{9(1 + t^2)}$$

$$Y = \frac{bt[t^3 - 3t + 2t_0(3 + t_0^2)]}{9(1 + t^2)}$$

Taking a full focal plane paraboloid $t_0 = -1$, then values $t = 0$ and $t = \pm 1$ encompass the entire derived part of the caustic (not essentially the best choice

32

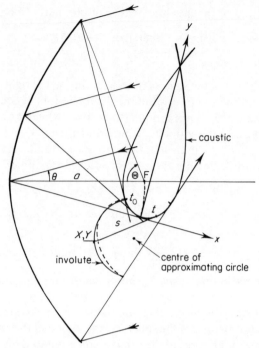

Figure 2.22 Approximation of the involute of Tschirnhausen's cubic by a circle

but the simplest). The three points lie on a circle with centre (in the oblique coordinate system)

$$(b/21, -11b/126)$$

In these examples we have, of course, found a general phase front, since involutes are parallel curves of the zero-distance phase front and the approximations to each will have the same centre.

2.6.2 Caustic matching

It is quite apparent from the foregoing that those systems which give a caustic where a true point focus is desirable, would, by a reversal of the rays, give the correct pattern of rays if the caustic were used as the source instead of the point source. Thus it is possible to use one system with a caustic approximately close to a second system with a caustic to give a corrected ray system. This can be illustrated by designing the well-known corrector for the spherical aberration of a circular mirror, the Schmidt corrector.[15] The caustic that requires to be matched is again the nephroid given by equations 2.24 and the matching caustic is to be that of a *plane* refraction from a source *inside* the refracting medium (Figure 2.23). In this case the zero-distance phase front is the ellipse found in section 2.1

$$(x - c)^2 - y^2/(\eta^2 - 1) = c^2/\eta^2 \qquad (2.26)$$

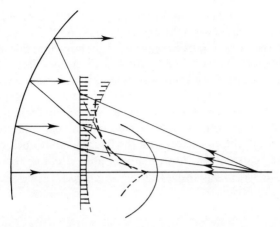

Figure 2.23 Matching the caustics of a reflector and a refracting surface. The caustic of the circular reflector can be matched approximately (paraxially) by the involute of the elliptical phase front of the refraction in a plane interface

The evolute of the ellipse

$$x = a \cos t$$

$$y = b \sin t$$

is, by equations 2.21

$$X = (a^2 - b^2)(\cos^3 t)/a$$
$$Y = (b^2 - a^2)(\sin^3 t)/a \qquad a > b \tag{2.27}$$

On relating this to the elliptical zero-distance phase front of equation 2.26, the caustic of the plane refraction is

$$X = c\eta(\cos^3 t)$$
$$Y = c\eta(\sin^3 t)(1 - \eta^2)^{-\frac{1}{2}} \tag{2.28}$$

with $\eta < 1$ through the choice of refraction from a source *within* the medium.

This is an 'elliptical' form of the astroid given by

$$X^{\frac{2}{3}}/C^{\frac{2}{3}} + Y^{\frac{2}{3}}/D^{\frac{2}{3}} = 1$$

The caustic of the reflection in the circle is the nephroid

$$X = a(6 \cos t - 4 \cos^3 t)$$

$$Y = 4a \sin^3 t$$

the same nephroid as equation 2.24 adapted. On reflecting this and translating it to match up with the refracting caustic, we have

$$X = 4a - 6a \cos t + 4a \cos^3 t = 2a \cos^3 t$$
$$Y = \qquad\qquad\qquad\qquad = 4a \sin^3 t \tag{2.29}$$

up to powers of t^4 or higher. Thus, by comparison with equation 2.28, not only do we derive a fair match between the curves of equations 2.28 and 2.29 but we find that the refractive index required is $\eta = \sqrt{3}/2$, inverted for real values to $2/\sqrt{3}$. It remains to obtain a method whereby the source can be effectively embedded in the medium, the reason for η being less than unity. This can be achieved by any second surface for which the source is aplanatic and most simply by a circular interface with the source at the centre, as shown in Figure 2.21. Thus the correction required for spherical aberration in a circular reflector derives from the plane surface of the corrector, with the figured surface being comparatively arbitrary. This figuring can, however, take into account the approximation remaining in the use of equation 2.29.

2.6.3 Progression of surfaces—paraxial approximation

The zero-distance phase front of a source and surface combination becomes the source itself in the image space of the surface and behaves as if in the infinite medium of the image space. In the presence of a second surface there is, therefore, a new source and surface combination, the zero-distance phase front of which can be obtained by the envelope method for extended sources. This will give the zero-distance phase front in the image space of the second surface and the procedure can be continued, by obvious means, through an entire system of surfaces involving both refracting and reflecting elements. At any stage the actual caustic of the rays can be derived by taking the evolute of the phase front at that stage.

From the complicated form that a zero-distance phase front may have, even after a single refraction from a point source, it is quite obvious that geometrically a cascading process will rapidly become unwieldy. This is mainly because the procedure as previously stated applies to the wide angle of rays issuing from the source. In the paraxial approximation only those regions of phase fronts about the axial direction would need to be considered. Approximations can then be made to those parts of the curve that would simplify the envelope method or provide a simpler method of progressing to the image space of the subsequent surface.

The process is illustrated in Figure 2.24 for a bi-convex spherical lens. For the first surface and source point S, the zero-distance phase front is a curve A, given by equation 2.6 with the appropriate values of k, η and r. The zero-distance phase front of A, now *interior* to the medium with the circular second surface, is obtained *exactly* as the envelope of zero-distance phase fronts B as a source point P moves over the curve A. Each separate curve B can be obtained from equations 2.6 or 2.7 with a parameter introduced to define the position of P on A. Elimination of this parameter by the standard procedure gives the envelope of curve B. The exact phase front in image space is then a parallel surface C to the envelope B, and the caustic the evolute of C.

Now surface A can be approximated by a circle in a small region about the axis. This makes it more probable that the envelope derived from the varying point P

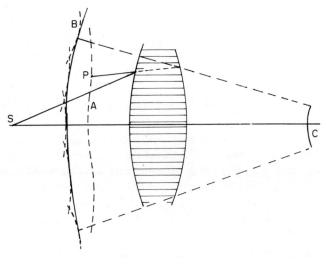

Figure 2.24 Phase front of a spherical lens. The zero-distance phase front of source and first surface is A. Points P on A give rise to zero-distance phase fronts of the second surface B. The envelope of B is produced to a parallel curve C. The evolute of C will be the caustic of the lens. In an approximation A and the envelope of B can be made circular in a paraxial region

can be obtained analytically. A spherical aberration (and higher symmetrical orders) has been introduced by this approximation.

For any subsequent spherical surface the envelope so derived has itself to be approximated by a circle. With judicious choice of approximating circles in both cases some cancellation of the introduced approximation aberrations could be achieved.

3 The Inversion Theorem of Damien

A major reason for establishing, in previous sections, a particular phase front, the phase front of zero distance, and examining its geometry, is that a unique property pertains to that phase front. This property was first explicitly enunciated by Damien[16] in a small monograph, although there are indications that a theorem of a similar nature was in the mind of Hamilton, and he refers to it as a continuation of a study by Cauchy. In its basic form the inversion theorem states (refer to Figure 3.1):

> Given a surface g and its zero-distance phase front h with respect to a source S, then an inversion in a circle centred on S will invert g into g' and h into h', such that g' is the zero-distance phase front of h' with respect to a source at S and for the same refractive index.

It applies both to refracting and reflecting surfaces. It is a most unusual transformation in that it transforms a phase front, fixed in time by making it of zero distance, into a refracting surface which is fixed in space. It is one of the very rare transformations that exist that completely transforms a given optical system, the first refraction, into a second optical system, purely geometrically. That it relies entirely on inversion is also relevant to the transformations already noted in section 2.5.2.

Damien himself gives a purely geometrical proof based on the similarity of the quadrilaterals SPMT and SM'P'T' in Figure 3.2. As can be seen the points P and M are transposed in this description making a refracting point M into a phase front point P. The proof given here is due to M. C. Jones.[17]

Refer to Figure 3.2 and put $\eta_2/\eta_1 = 1/m > 1$. The construction of the zero-distance phase front gives

$$(1 - m^2)r^2 - 2rr' \cos (\gamma - \gamma') + r'^2 = 0 \tag{3.1}$$

Snell's law of refraction at M gives

$$(1 - m^2)r(\cos \gamma + \sin \gamma \tan \psi) = r'(\cos \gamma' + \sin \gamma' \tan \psi) \tag{3.2}$$

Also

$$\frac{dr}{r d\gamma} = \frac{\cos \gamma + \sin \gamma \tan \psi}{\cos \gamma \tan \psi - \sin \gamma}$$

$$= \frac{r' \sin (\gamma' - \gamma)}{r(1 + m^2) - r' \cos (\gamma - \gamma')} \tag{3.3}$$

from equation 3.2.

36

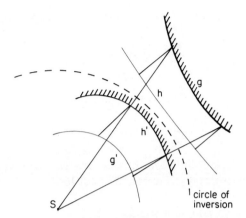

Figure 3.1 The inversion theorem of Damien. h is the zero-distance phase front of g. After inversion g′ is the zero-distance phase front of h′

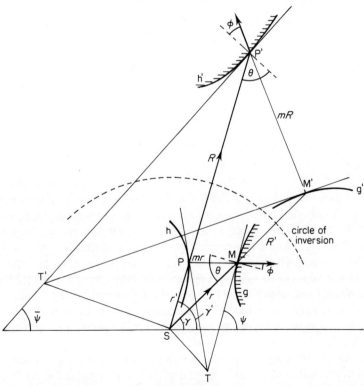

Figure 3.2 Proof of the inversion theorem. Note that the quadrilaterals SPMT and SM′P′T′ are similar

On differentiating equation 3.1 with respect to γ and applying equation 3.3, we obtain

$$[r\cos(\gamma - \gamma') - r']\frac{dr'}{d\gamma} + rr'\sin(\gamma - \gamma') = 0 \qquad (3.4)$$

Equations 3.1 and 3.4 are required to be invariant under the inversion

$$r \rightarrow \frac{1}{R'} \qquad r' \rightarrow \frac{1}{R} \qquad \gamma \leftrightarrow \gamma'$$

that is primed (phase-front) coordinates are transformed into unprimed (refractive) coordinates, and vice versa. Without loss of generality the radius of inversion has been taken to be unity.

Equation 3.1 is invariant as it stands.

For the refraction at P' we must have (cf. equation 3.2)

$$(1 - m^2)R(\cos\gamma' + \sin\gamma'\tan\bar{\psi}) = R'(\cos\gamma + \sin\gamma\tan\bar{\psi}) \qquad (3.5)$$

where

$$\tan\bar{\psi} = \left(\cos\gamma' + \sin\gamma'\frac{dR}{Rd\gamma'}\right)\bigg/\left(-\sin\gamma' + \cos\gamma'\frac{dR}{Rd\gamma'}\right)$$

$$= \left(-\cos\gamma + \sin\gamma\frac{dR'}{R'd\gamma}\right)\bigg/\left(\sin\gamma + \cos\gamma\frac{dR'}{R'd\gamma}\right)$$

On substituting this result into equation 3.5 and transforming the resultant equation, we have

$$(1 - m^2)\frac{rdr'}{r'd\gamma'} = \sin(\gamma' - \gamma) + \cos(\gamma' - \gamma)\frac{1}{r'}\frac{dr'}{d\gamma'}$$

which, using equation 3.1, reproduces equation 3.4.

Damien's theorem can, in an obvious way, be used to obtain the zero-distance phase front of a surface from the known zero-distance phase front of another surface. Thus for reflection in a circle, the situation considered in section 2.3, we require a reflector that has a known circular zero-distance phase front. For this we take the ellipse with source at one focus (Figure 3.3), for which the zero-distance phase front is a circle centred on the second focus. The source is thus interior to the circle. The inversion in a circle centred on the source transforms the circular phase front into the circular reflector, and the elliptical reflector into the required phase front, which is the limaçon of Pascal without cusp from the table in Appendix II. Variation of the radius of inversion or utilization of the concept of negative inversion provides for the other situations of source and reflector given in section 2.3.

It is of interest to compare this result with the symmetry between surfaces and wavefronts illustrated at the end of section 2.2. The two instances, (a) and (b) of Figure 2.9, can be seen to be the result of an inversion, if the centre of inversion is taken to be the source at infinity, that is, a reflection. The limaçon and (eccentric)

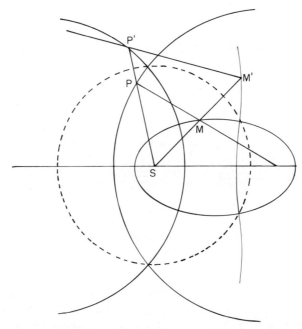

Figure 3.3 Application of the inversion theorem to derivation of zero-distance phase fronts

circle combination shown in Figure 2.9(c), is a direct application of Damien's theorem to the ellipse combination in Figure 2.9(b), with the centre of inversion this time at the source at the focus of the corresponding ellipse. Thus the ellipses in each case invert into limaçons of Pascal, and the plane surfaces into spheres.

3.1 A GEOMETRICAL METHOD OF OPTICAL DESIGN

We consider now the case where a source and the first surface of any optical system are given and it is required to determine the appropriate second surface that will bring all the rays to an exact specified focus. This will in principle always be possible for rays issuing from a point source after a finite number of reflections or refractions.

We can deal in the first instance with the design of the second surface of a lens for which the first surface, the source and the required image are specified.[17] Then, as in Figure 3.4, the zero-distance phase front A of the source and first surface can be derived by any of the methods so far given. This phase front replaces completely both the source and the first surface and we regard it as a new source in the infinite medium with the given refractive index. We form at an arbitrary distance the parallel A' of the curve A and we will show that this arbitrary choice is solely dependent on the so far arbitrary thickness of the lens at its centre.

In order to obtain the required focusing action the surface A' must be the zero-

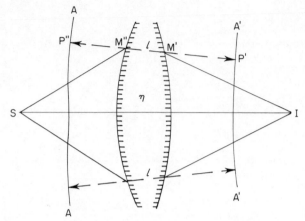

Figure 3.4 Transmission through a perfect lens. A is the zero-distance phase front of the source and first surface, A' that of the image and second surface. A' is a parallel of A

distance phase front of some refracting surface with respect to the specified image I. For then, in the literal meaning of zero phase, this occurs between the source and the surface A, and also between the image and the surface A'. Since A and A' are parallel in a uniform medium, the phase along mutual normals is constant for all rays, and thus the total phase from source to image is constant along all rays. Hence the image will be a perfect focus.

What is required to complete the design is the process whereby the second surface is derived from the now known phase front A'. This process is carried out in three stages (refer to Figure 3.5):

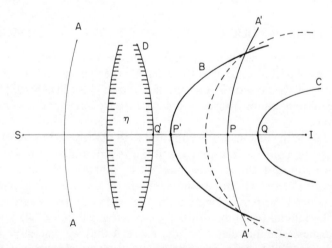

Figure 3.5 Construction of the second surface. A is the zero-distance phase front of the source and first surface, A' is a parallel. B is the inverse of A' in a circle centred on the image. C is the zero-distance phase front of B which inverts into D, the required second surface

(a) Invert the surface A' in a circle (of arbitrary but appropriate radius) centred on I to give curve B.

(b) Obtain the zero-distance phase front of B with respect of I with the refractive index of the medium concerned to give curve C.

(c) Invert curve C in the same circle as in (a) to give curve D.

Curve D is the profile of the second surface required. If curve A' intersects the axis SI at P, curve B at P', curve C at Q and curve D at Q' as shown, then the position of P is solely dependent on the chosen axial thickness of the lens and the process is independent of the radius of inversion. We have by the construction of zero-distance phase fronts

$$Q'P = Q'I/\eta \qquad (3.6)$$

or

$$IP = (1 - 1/\eta)IQ'$$

P' is derived by an inversion, so $IP \cdot IP' = \kappa^2$ (κ being the radius of inversion); Q is derived by the zero-distance phase front procedure

$$P'Q = P'I/\eta$$

or

$$IQ = (1 - 1/\eta)P'I$$

and Q' is derived from Q by an inversion of the same radius $IQ \cdot IQ' = \kappa^2$. Hence

$$IQ' = \frac{\kappa^2}{IQ} = \frac{\kappa^2}{(1 - 1/\eta)IP'} = \frac{\kappa^2}{(1 - 1/\eta)} \frac{IP}{\kappa^2}$$

which is identical to equation 3.6 and independent of κ^2.

The complex nature of the geometrical curves that arise in this procedure makes demonstrations of the method difficult in all but a few well-known optical designs. Of these we take the astigmatic lens with hyperbolic surfaces and the Gregorian reflector of the parabolic mirror. For these cases the zero-distance phase fronts of the first surface and source are plane and circular respectively.

For the former we are given the source S at the focus of a hyperbola, the interface with a medium of refractive index η (Figure 3.6). If the asymptotes of the hyperbola are at an angle $\pm\theta$ such that $\theta = 1/\eta$, all the rays are refracted to become parallel to the axis. The zero-distance phase front is therefore a plane A, perpendicular to the axis and its parallel similarly a plane A' perpendicular to the axis. The inversion of A' in a circle centred at I is therefore the circle B passing through I in Figure 3.6. The refractive zero-distance phase front of a circle with a source on its perimeter ($r = k$ in equation 2.6) is the interior loop of a limaçon with cusp at I on curve C. It is simple to show that tangents to the cusp make the same angle θ with the axis at I, and hence the inversion of the cusped curve will have asymptotes making the same angle with the axis. The inversion of a limaçon being a conic with asymptotes is thus the second hyperbolic surface D producing an exact focus at I.

For the parabolic reflector, as we have shown, the zero-distance phase front is a

42

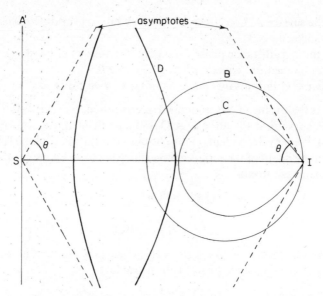

Figure 3.6 Application of the method to the exact lens with hyperbolic surfaces

circle A, strictly of infinite radius, centred on the focus F. Its parallel is therefore any other circle A′ with appropriate radius also centred on F (Figure 3.7). We require the second surface to create a focus at an arbitrary point, which for simplicity we have chosen to be the apex I of the parabola. The inversion of A′ in a circle centred on I is a circle B as shown and the *reflective* zero-distance phase front of B is the limaçon C. The inversion of the single-loop limaçon in its pole is the ellipse with focus at the pole. Hence the second required reflector is the ellipse with I and F as foci. The form of this conic can be varied in many ways by the choice of the radius of A′, the radius of inversion and the sign of the inversion. This is possible since the eccentricity of the ellipse is arbitrary up to the degree that converts it into a hyperbola giving the Cassegrain form of sub-reflector. In all cases I and F become the two foci in this process.

The converse of the method also has applications for deriving the zero-distance phase fronts in situations where other methods might be too complicated. Thus if we have a two-surfaced system with known exact focusing properties, any one of the stages unknown in the procedure can be derived from the others which are known. For example, the two-surface lens with a circular first surface centred on the source and an elliptical second surface with focus at the source with eccentricity $\varepsilon = 1/\eta$ collimates all rays parallel to the axis (Figure 3.8). The zero-distance phase front of the first surface is the circle A centred on S, which we can take also to be its parallel A′ as an arbitrary choice of inversion arises later. The inversion of A′ in the point at infinity is a reflected circle B wherewith we take up the arbitrariness by making B tangential to the first surface of the lens and of radius SS′. The zero-distance phase front of B must be the curve C which will *reflect* (inverse at infinity) into the known second surface D. It is therefore itself

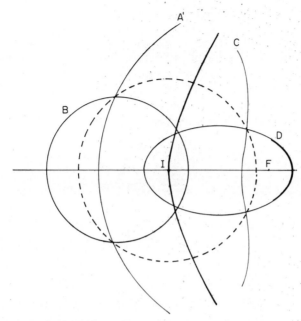

Figure 3.7 Application of the method to the Gregorian reflector

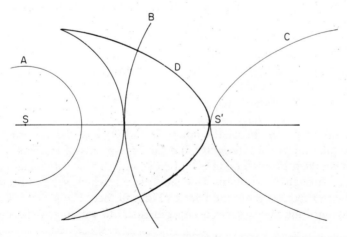

Figure 3.8 The converse of the method for a known two-surface lens, giving an unknown zero-distance phase front

the reflection of D. Hence the zero-distance phase front of internal refraction of parallel rays at a circular interface is the ellipse with eccentricity $\varepsilon = 1/\eta$. This can be proved to be the case by the method of elimination of the parameter of a moving transverse source as in section 2.4.

3.2 LENS BENDING

The final example is an illustration of an iterative procedure based on the method of inversions which is in principle a 'lens-bending' process to produce to any required degree of accuracy a lens with specified properties. The lens specified in this instance is to collimate rays from a source to rays parallel to the axis (of symmetry) *and* to be symmetrical about a plane perpendicular to that axis (Figure 3.9). Approximations to this lens have been made in the form of a corrected plane

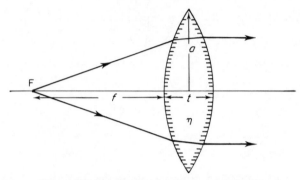

Figure 3.9 Parameters for the symmetrical lens

reflector, which is obtained from a perfectly symmetrical lens by inserting a reflecting surface in the plane of symmetry.[18] We can therefore choose any appropriate, obviously convex first surface. Now a *symmetrical* lens of specified diameter, focus and refractive index will automatically have its axial points determined by the equality of ray paths through the centre and through the edge. Hence in Figure 3.9 we must have

$$t = [(f^2 + a^2)^{\frac{1}{2}} - f]/(\eta - 1)$$

thus determining t and hence the points O and O'. For a first surface we can therefore take the circle through O and the edge points of the lens. Using the procedure of the previous section to find the second surface with image at infinity that passes through O', we obtain a second surface D which is then reflected in the mid-plane to become the second choice of first surface. The process is repeated iteratively until the two surfaces become identical to within any specified limit.

3.3 EXTENDED SOURCES AND IMAGES

There are many situations in practice where an extended source or image field is required instead of either the perfect focus or ideal point source to which the inversion theorem applies. In these conditions the off-axis asymmetry necessarily implies an approximation procedure. In optical design theory these create aberrations and the cancellation of these by additional refracting surfaces is the guiding principle of the design. The fact that a plurality of phase fronts can be

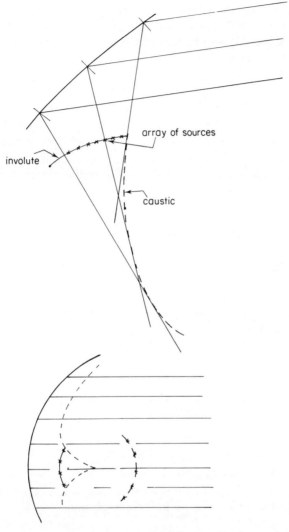

Figure 3.10 An array of sources on the involute of the caustic creates the ray pattern of the caustic

represented approximately by their envelope, when this exists, means this envelope can be applied to the methods resulting from the zero-distance phase front inversion shown previously.

3.3.1 Multiple point sources

The reciprocity between the caustic created by a reflector and an incident beam of radiation, means that the creation of the caustic as a source would reproduce the beam as outgoing radiation. A means of creating such a source is to develop the involute of the caustic and regard it as the source phase front. Point sources at an appropriate spacing (usually less than $\lambda/2$ to avoid grating lobe effects) placed along this involute would then, by virtue of Huygens' construction, create the necessary parallels giving the caustic in one direction and the required radiating beam in the other (Figure 3.10).

If instead of a single beam and single caustic, several were to be used with radiation incident over an angular field, the separate involutes would have an envelope which would thus become the base line for an array of sources, or a moving single source, permitting the scanning of the beam over the angular range (Figure 3.11).

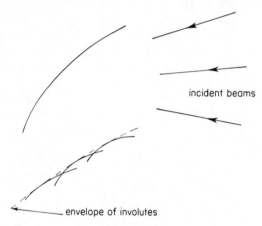

incident beams

envelope of involutes

Figure 3.11 Base line for the array for many beams is the envelope of the involutes of the individual beams

By the same process a shaped radiation pattern can be created by reducing it to individual beams of specified amplitude and phase and utilizing the individual caustics through the envelope of their involutes. This same method can be used with the inversion theorem as shown in the next section.

3.3.2 Shaped-beam antennas

The problem to be approached is that of curved single or double reflector systems with point or extended sources giving shaped radiation patterns. As is well known this problem is complicated by indeterminacy of phase. The process to be outlined separates the effects of phase and amplitude in a way that appears to make iterative or optimization techniques particularly simple.

In Figure 3.12 the required radiation pattern is reduced to a basic minimum of plane waves incident upon the given surface YY' and at different incidence angles. The zero-distance phase fronts for these waves are then derived by the process for an infinite source giving the curves a, b, c, . . . each of which will be different in form owing to the difference in incidence angle. Each will have a phase datum, say

47

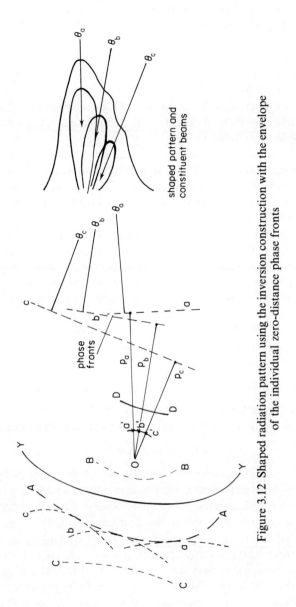

Figure 3.12 Shaped radiation pattern using the inversion construction with the envelope of the individual zero-distance phase fronts

the distance of the perpendicular p, from a central point such as O (not necessarily a focus but generally some obvious mean position).

The envelope of these curves is the curve A originating the procedure of section 3.1. Now it can easily be seen that the phase of each wave can be individually adjusted by the length of the appropriate perpendicular from O. But this selection cannot be totally arbitrary as the requirement that the resultant set of curves must have an envelope has to be obeyed.

The curve A shown in Figure 3.12 is that assumed to be given by a continuous decrease in p from a to b, and so on.

Curve A is now inverted to the image point O resulting in curve B as before. Its zero-distance reflective phase front is C, and the inversion of C gives the required shape of the second reflector D.

We now surround O with a small circle which will be the distribution of the fundamental sources a′, b′, c′, ... To each of these sources a plane wave can be attributed through the geometry of the procedure. Hence, an amplitude and phase distribution can be imposed on the sources equal to that required by the original spectrum of waves.

This can be generalized in an obvious manner for sources distributed over other surfaces, say a planar phased array. If it is repeated in the other planes of a two-dimensional shaped pattern, the resulting curves construct a surface for full pattern shaping. A single-surface reflector can be obtained by the same method if YY′ is considered in the first instance to be a hypothetical infinite plane.

The process thus fulfils the fundamental role of geometrical optics by providing a sufficiency of designs which can be assessed or optimized by the computational methods of physical optics or diffraction theory.

3.3.3 The sine condition

The Abbe sine condition for a displacement of the source by an amount dl perpendicular to the axis, requires, with source and focus both in free space, that

$$dl \sin \theta = dl' \sin \theta' \tag{3.7}$$

where θ is the semi-angle subtended by the aperture at the source and primes refer to image space. The inversion process of section 3.1 was specifically based on the arbitrary choice of the first surface. Since the addition of a second condition for a two-surface system, the first being the focusing property itself, completely specifies *both* surfaces, thus an *a priori* choice cannot be made. There appears to be no method whereby the inversion process gives a feedback or iterative process to adjust the first surface so that the eventual second surface will give a system satisfying the sine condition. This must remain as an as yet unsolved problem with this technique. The only procedure thus available is to repeat the design process with a variety of choice of the first surface until the criterion is adequately met.

4 The Mechanical Description of Optical Surfaces

4.1 THE CO-INVOLUTION OF TWO CAUSTICS

The process of drawing the involute of a curve by a string wound about the curve was given in section 2.5.3. We now propose an extension of this method involving two caustics (Figure 4.1), in which a taut inextensible string is fixed to points A_1 and B_1 on the caustics A and B enveloping the curves and held taut by the drawing point P as the string unwinds from one curve and winds onto the other. The locus of P is a curve C, the reflecting curve which will reflect all the rays creating caustic A into the envelope creating caustic B. If caustics A and B were to degenerate to perfect point foci, the method of drawing the reflector is the well-known method for drawing an ellipse with a string stretched over two pins at the foci. This construction was given by Leibniz in a letter to John Bernoulli in 1704.[19] The curves can be wound or unwound in the same or in contra directions, as shown in Figure 4.1. The proof that C is a reflector with this property arises from an application of Fermat's principle. If, as in Figure 4.2, A and B are points where the taut string is tangential to the caustics, then an incremental movement ds of the drawing point will 'wind on' Δr on the A caustic and 'unwind' $\Delta\rho$ on the B caustic, and obviously Δr will be equal in length to $\Delta\rho$. Hence from the figure we will have, measuring r and ρ from the intersection of the tangents from C_1 and C_2 respectively

$$r + \delta r + \rho - \delta\rho = \rho + d\rho + \delta\rho + r + dr - \delta r$$

Since $\Delta\rho = \Delta r$, and to the first order $\Delta\rho = 2\delta\rho$ and $\Delta r = 2\delta r$, we have

$$dr = -d\rho$$

which from Chapter 1 defines a reflector obeying Snell's law at C.

The term proposed by Leibniz for this operation was 'co-ëvoluteo', but owing to its similarity to the drawing of involutions of curves, the term 'co-involution' would seem more appropriate.

We shall describe the method by the coordinate geometry of the caustics. Let caustic A have parametric description $x = f_1(t)$, $y = g_1(t)$, and caustic B parametric description $x = f_2(s), y = g_2(s)$; then a point (X, Y) on curve C obeys

$$[(X - f_1)^2 + (Y - f_1)^2]^{\frac{1}{2}} + \int_{t_0}^{t} (f_1'^2 + g_1'^2)^{\frac{1}{2}} dt$$
$$+ [(X - f_2)^2 + (Y - g_2)^2]^{\frac{1}{2}} + \int_{s_0}^{s} (f_2'^2 + g_2'^2)^{\frac{1}{2}} ds = L \quad (4.1)$$

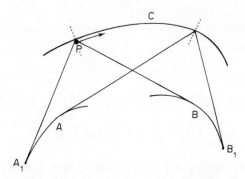

Figure 4.1 Co-involution of two caustics A and B. The point P is held in a taut string wound over the two curves

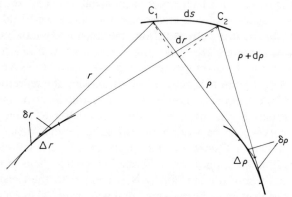

Figure 4.2 Proof that the co-involution drawing process gives rise to a reflecting surface C

where L is the fixed length of string and t_0 and s_0 are the parameters of the points A_1 and B_1 where the string is attached to each caustic. Since the string is a tangent to A at the point with parameter t and to B at s, we have in addition

$$\frac{Y - g_1(t)}{X - f_1(t)} = \frac{df_1}{dg_1} \tag{4.2a}$$

$$\frac{Y - g_2(s)}{X - f_2(s)} = \frac{df_2}{dg_2} \tag{4.2b}$$

If the caustic B is reduced to a point focus at the point (a, b) the same equations apply with $f_2 = a$, $g_2 = b$ and the second line integral in equation 4.1 and equation 4.2b can be omitted.

4.1.1 Sub-reflector for parabola with offset beam

The caustic of a parabola $y^2 = 4ax$ receiving a beam of parallel rays making an angle θ with the axis is the cubic already described in section 2.6.1,

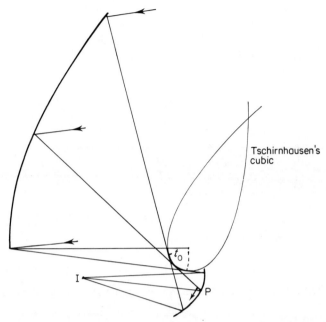

Figure 4.3 Drawing method for a sub-reflector in an asymmetrically illuminated parabola by a string wound about Tschirnhausen's cubic

Tschirnhausen's cubic (Figure 4.3). In the coordinate system of the paraboloid itself this has equations

$$x = \{-at \sin \theta(t^2 - 3) + a \cos \theta\} \cos \theta + 3at^2 \sin^2 \theta \equiv f_1(t)$$

$$y = \{at \sin \theta(t^2 - 3) - a \cos \theta\} \sin \theta + 3at^2 \cos^2 \theta \equiv g_1(t)$$

To produce a focus at the origin, put $a = 0$, $b = 0$ in equations 4.1 and (dropping the suffix since the second caustic is a point focus)

$$S(t) \equiv \int_{t_0}^{t} \{f'^2 + g'^2\}^{\frac{1}{2}} dt = at \sin \theta(t^2 + 3) - at_0 \sin \theta(t_0^2 + 3) \qquad (4.3)$$

t_0 is the point of fixing on the caustic, which for a paraboloid subtending a total angle of 2Θ is given approximately by

$$t_0 = \tan (\Theta/2)$$

Substitution into equations 4.1 gives the implicit parametric equations

$$S(t) + (Y - g)(1 + g'^2/f'^2)^{\frac{1}{2}} + \{[(Y - g)g'/f' + f]^2 + Y^2\}^{\frac{1}{2}} = L$$

$$S(t) + (X - f)(1 + f'^2/g'^2)^{\frac{1}{2}} + \{[(X - f)f'/g' + g]^2 + X^2\}^{\frac{1}{2}} = L$$

with

$$f'/g' = [(1 - t^2) \sin \theta \cos \theta + 2t \sin^2 \theta]/[(t^2 - 1) \sin^2 \theta + 2t \cos^2 \theta]$$

$$(4.4)$$

The range of t in these equations is $\pm t_0$.

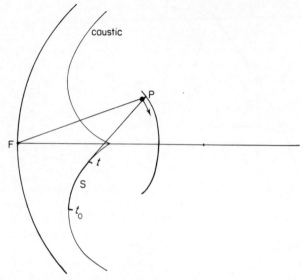

Figure 4.4 The corrector for a spherical mirror with a string wound about the nephroid caustic

4.1.2 Correctors for spherical mirrors

The technique for obtaining the symmetrically placed sub-reflector for correcting the spherical aberration of a spherical mirror is illustrated in Figure 4.4. The caustic is the nephroid and the relevant equations are those given in section 2.6.1 adapted and used in equation 4.1. Hence

$$f(t) = a(3 \cos t - \cos 3t)$$

$$g(t) = a(3 \sin t - \sin 3t) \qquad -\pi < t < \pi$$

The cusp is at the value $t = -\pi$ as drawn in Figure 4.4. The integral of equation 4.3 is then

$$S(t) = -6a \cos t \qquad f'/g' = -\tan 2t$$

These values can be substituted into equations 4.1 to give, for example, a focus at the apex $(-4a, 0)$.

This case is of particular interest since an offset sub-reflector can be designed in an identical manner to give an axial beam not disturbed by the blockage of the centrally placed sub-reflector. This is shown in Figure 4.5.

4.1.3 The cardioid reflector

The caustic of the circular reflector which has been used continuously in these studies appears again in the application of co-involution to the circle–cardioid combination known as the Zeiss Cardioid.[8] By its construction the circular reflector is the co-involution of a nephroid caustic at a point at infinity. But, from

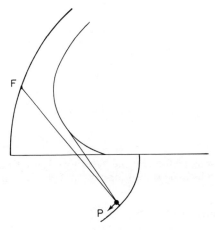

Figure 4.5 Arrangement for elimination of blockage in a circular reflector

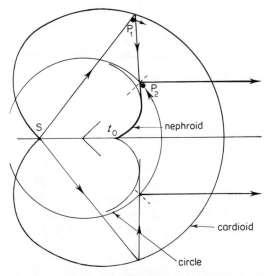

Figure 4.6 The co-involution of the caustic of the circle with the point at infinity gives the circle, and the co-involution of the source and the caustic gives the cardioid. This is then the Zeiss Cardioid combination

the table in Appendix II.3 the nephroid is also the caustic by reflection of a point source at the cusp of a cardioid; thus, as shown in Figure 4.6, the cardioid will focus all the rays previously reflected by the circle at its cusp.

4.1.4 Reflectors for virtual caustics

A similar taut string technique applies to the description of a hyperbola when related to point foci. The image is then a virtual focus and the curve is the solution

54

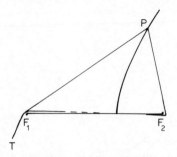

Figure 4.7 A string drawing method for the hyperbola. P is a fixed point on the string looped over the two foci and drawn together at T. F_2 is a virtual focus of F_1

of equation 1.1 with positive sign, or

$$r - \rho = \text{constant} \qquad (4.5)$$

This curve can be described by a *fixed* point P on a string which is looped over the two foci F_1 and F_2 as shown in Figure 4.7, going on to a combined point T at a sufficient distance. If the combined strings move in the direction T, both parts F_1P and F_2P are reduced at the same rate and hence the difference

$$F_1P - F_2P = \text{constant} = r - \rho$$

satisfies equation 4.5.

By a similar analysis to that in section 4.1.1, the point caustics F_1 and F_2 can be replaced by line caustics and the string overlaid in the same manner as in Figure 4.1. This gives a reflector that transforms the real caustic at A into a virtual caustic at B, as illustrated in Figure 4.8. This method could be used as shown in Figure 4.9 to obtain the Cassegrain design of sub-reflector that would convert the caustic, the Tschirnhausen cubic discussed in section 2.6.1, to an arbitrarily situated focus. The string is combined over this required focus, one branch going to the fixed point P and the other winding about the caustic and looping over the second

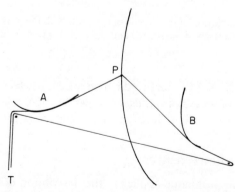

Figure 4.8 Extension of the co-involution process to a virtual and a real caustic

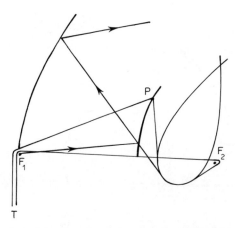

Figure 4.9 Application of the virtual caustic process to the design of a Cassegrain reflector with off-axis beam

focus F_2 on that caustic. Note that only one part of the string at the focus F_2 has to be wound over the caustic to retain the validity of Fermat's principle.

4.2 THE GENERAL TWO-SURFACE REFLECTOR SYSTEM

Ideally the method could be made fully complete if there were a mechanical drawing method by which the original caustic could be derived. This lack can be avoided by a more general procedure. In the illustrative examples the common basic reflectors were taken, namely the circle and the parabola, each of which has a fairly complex caustic. However, it is obvious that we could instead choose a completely arbitrary caustic of simple definition, form a co-involution of it with one specified focus giving a primary reflector and a second co-involution with a second specified focus giving a sub-reflector. The two-reflector combination would then focus the first source into the second image. In this manner we trade simple reflectors with complex caustics for simple caustics with not too complex reflectors. Taking, for example, a circular caustic and a source at infinity, as in Figure 4.10, the two co-involutions performed simultaneously describe the main and subsidiary reflectors. The process can be termed simultaneous double co-involution. As is now obvious from this description the tangents to the caustic in both co-involution processes are contiguous. The parts of the tangents for which this is so could then be taken by a taut string contained within a rigid but extendable tube. As the tube rolls without slipping about the caustic, and the string is kept taut, the ends simultaneously describe both the main and subsidiary reflectors, a method illustrated in Figure 4.11. By rolling a link of this sort about a point caustic, the simultaneous drawing of the parabola and Gregorian elliptical sub-reflector can be shown to be an extension of the simple ellipse drawing method alone. It is possible to develop the method further and have several tubular links of the kind shown, each rolling without slipping on a specified

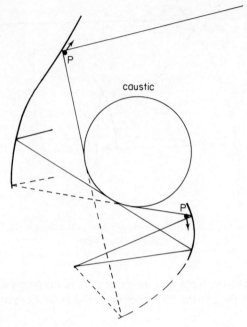

Figure 4.10 Simultaneous co-involution of a primary and a secondary reflector from an arbitrary mutual caustic

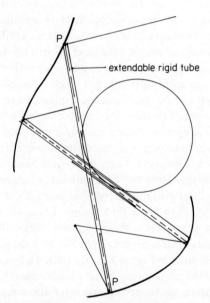

Figure 4.11 Simultaneous double co-involution, the tangential part of the string held in a rigid extendable tube

caustic curve to give the profiles of a multiple reflector system with final end focusing requirements. Such a procedure would be a simultaneous multiple co-involution. The process extends naturally to the two-dimensional design of reflector surfaces by the rolling of a tubular link over a caustic surface, provided only that the parts of the string external to the tube and that part inside the tube are all kept coplanar.

4.3 CO-INVOLUTION BY REFRACTION

A string drawing technique, suitable for adaptation to the process of co-involution, is known, for the point focusing by a single refraction as occurs with the Cartesian ovals given in Chapter 1.

As shown in Figure 4.12, if the string is looped one additional turn about one of the two foci, the point P, sliding inside the string, will be subject to the law

$$F_1P + 2F_2P = \text{constant} = L \text{ (the length of the string)} \qquad (4.6)$$

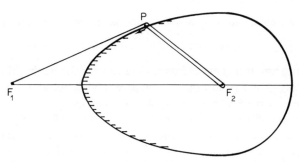

Figure 4.12 Taut string drawing process for a refraction with refractive index 2

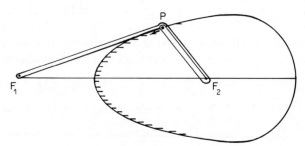

Figure 4.13 Drawing process for refractive index 3/2

This is the direct bipolar equation of a Cartesian oval from $r + 2\rho = \text{constant}$, implying a refractive index of two. Additional loops about the two foci apply in the same way to the derivation of the surface for any whole-number-fraction refractive index. That for $\eta = 1.5$ is shown in Figure 4.13.

We can extend this to the process of co-involution by winding or unwinding a

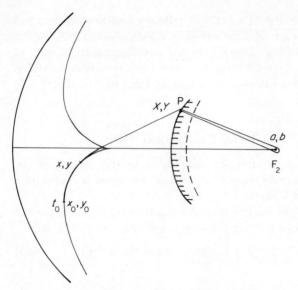

Figure 4.14 Co-involution with refraction giving the Schmidt corrector profile for a circular reflector

part of the string about a caustic. Performing this for the nephroid caustic of a circular reflector, as shown in Figure 4.14, gives rise directly to the mechanical drawing of the Schmidt corrector for a spherical mirror.[22] Of course this produces a single-surface refraction, and hence the second focus F_2 is embedded within the denser medium. It is 'brought out' by any second refracting surface for which F_2 is an aplanatic point and most simply by a spherical surface with F_2 at its centre. The refractive index in the case illustrated is $\eta = 2$. Defining the surface in terms of algebraic geometry, however, allows the use of *any* given refractive index. The implicit equation to the surface states the fact that the string from (x_0, y_0) (Figure 4.14) to (x, y) and then tangentially to (X, Y) has to equal (the negative of) the distance from (X, Y) to the required focus (a, b) multiplied by the specified refractive index, to within a constant, the length of the string

$$S + [(X - x)^2 + (Y - y)^2]^{\frac{1}{2}} = L - \eta[(X - a)^2 + (Y - b)^2]^{\frac{1}{2}} \qquad (4.7)$$

where S is the arc length of the caustic from (x_0, y_0) to (x, y), with an additional tangential relation of the form given in equation 4.2a.

5 Ray-tracing in Non-uniform Media

The tracing of rays in an isotropic non-uniform refractive medium provides a rich field for the geometry of curves in space and the differential geometry of surfaces. This not only applies to the design of practical microwave optical devices, but, being based upon a least-action integral, can be associated with the trajectories of particles in non-uniform potential fields and through this to many other branches of physical science. Recently techniques have become available which make possible the production of non-uniform optical glasses, leading to the design of optical components for fibre-optics or integrated optical systems.

As in the previous chapters we find that the practicability of the basic design methods is underlined by a transformation process of great theoretical complexity, and these too are beginning to find applications in the other least-action problems in physics. To demonstrate this equivalence we use the derivation of Hamilton's canonical equations for rays given by Synge.[20]

5.1 THE RAY-TRACING EQUATIONS

For any curve defined parametrically by

$$x = x(u) \qquad y = y(u) \qquad z = z(u)$$

to be a geometrical ray, it must obey Fermat's principle, that its optical length

$$V = \int_{u_1}^{u_2} \eta(x, y, z) \left[\left(\frac{dx}{du} \right)^2 + \left(\frac{dy}{du} \right)^2 + \left(\frac{dz}{du} \right)^2 \right]^{\frac{1}{2}} du \qquad (5.1)$$

is an extremum. We term the integrand W for brevity. The variation in passing from one curve to another near neighbour is

$$\delta V = \int_{u_1}^{u_2} \left(\sum \frac{\partial W}{\partial \dot{x}} \delta \dot{x} + \sum \frac{\partial W}{\partial x} \delta x \right) du$$

$\sum$ indicating the sum of similar terms with x, y and z, and the dot representing differentiation with respect to u. Thus $\delta \dot{x} = d/du \, (\delta x)$ and integration by parts gives

$$\delta V = \left[\sum \frac{\partial W}{\partial \dot{x}} \delta x \right]_{u_1}^{u_2} - \int_{u_1}^{u_2} \sum \left[\frac{d}{du} \left(\frac{\partial W}{\partial \dot{x}} \right) - \frac{\partial W}{\partial x} \right] \delta x \, du = 0$$

The first part vanishes if the curves have common end-points, and thus the integrand in the second term must equal zero. For arbitrary displacements δx this implies the Euler–Lagrange equations for a ray

59

$$\frac{\mathrm{d}}{\mathrm{d}u}\left(\frac{\partial W}{\partial \dot{x}}\right) - \frac{\partial W}{\partial x} = 0$$

$$\frac{\mathrm{d}}{\mathrm{d}u}\left(\frac{\partial W}{\partial \dot{y}}\right) - \frac{\partial W}{\partial y} = 0 \qquad (5.2)$$

$$\frac{\mathrm{d}}{\mathrm{d}u}\left(\frac{\partial W}{\partial \dot{z}}\right) - \frac{\partial W}{\partial z} = 0$$

On substituting the expression for W we have

$$\frac{\mathrm{d}}{\mathrm{d}u}\left[\frac{\eta\dot{x}}{(\dot{x}^2 + \dot{y}^2 + \dot{z}^2)^{\frac{1}{2}}}\right] - \frac{\partial\eta}{\partial x}(\dot{x}^2 + \dot{y}^2 + \dot{z}^2)^{\frac{1}{2}} = 0$$

The choice of u as parameter is arbitrary and we can make two specific selections:

(a) $u = s$, the arc length along the curve. We then have three equations of the form

$$\frac{\mathrm{d}}{\mathrm{d}s}\left(\eta\frac{\mathrm{d}x}{\mathrm{d}s}\right) - \frac{\partial\eta}{\partial x} = 0$$

which collectively for terms in x, y and z are

$$\frac{\mathrm{d}}{\mathrm{d}s}\left(\eta\frac{\mathrm{d}\mathbf{r}}{\mathrm{d}s}\right) = \nabla\eta \qquad (5.3)$$

where $\mathbf{r}$ is a position vector of a point on the ray from a given origin.

(b) Alternatively selecting u to be $u = \int \mathrm{d}s/\eta$, then $\mathrm{d}u = \mathrm{d}s/\eta$ and the ray equations become

$$\frac{\mathrm{d}^2x}{\mathrm{d}u^2} = \frac{\partial}{\partial x}(\tfrac{1}{2}\eta^2)$$

$$\frac{\mathrm{d}^2y}{\mathrm{d}u^2} = \frac{\partial}{\partial y}(\tfrac{1}{2}\eta^2) \qquad (5.4)$$

$$\frac{\mathrm{d}^2z}{\mathrm{d}u^2} = \frac{\partial}{\partial z}(\tfrac{1}{2}\eta^2)$$

Equations 5.2, 5.3 and 5.4 are different forms of the equations of rays, of which equation 5.3 will be seen to be the most applicable.

We now consider a variation of V in which the end-points too are infinitesimally displaced, and we take the parameter u to equal s, the arc length along a ray. Then

$$\frac{\partial W}{\partial \dot{x}} = \eta\alpha \qquad \frac{\partial W}{\partial \dot{y}} = \eta\beta \qquad \frac{\partial W}{\partial \dot{z}} = \eta\gamma$$

where α, β and γ are the direction cosines of the ray at its starting point. Using primes for similar entities for the ray at its end-point, e.g. $\partial W'/\partial \dot{x}' = \eta'\alpha'$, we have

$$\delta V = \sum_{\substack{x,y,z \\ \alpha,\beta,\gamma}} \eta\alpha - \sum_{\substack{x',y',z' \\ \alpha',\beta',\gamma'}} \eta'\alpha'$$

and thus

$$\frac{\partial V}{\partial x} = \eta\alpha \qquad \frac{\partial V}{\partial y} = \eta\beta \qquad \frac{\partial V}{\partial z} = \eta\gamma \quad \text{etc.}$$

Since α, β and γ are direction cosines, this results in

$$\left(\frac{\partial V}{\partial x}\right)^2 + \left(\frac{\partial V}{\partial y}\right)^2 + \left(\frac{\partial V}{\partial z}\right)^2 = \eta^2 \tag{5.5}$$

with a similar result for the primed coordinate system.

Finally, for an isotropic medium, rays are normal to wavefronts, and we can use Huygens' construction. If $S(x, y, z) = 0$ is the equation of a wavefront reaching the point (x, y, z) at time t, then the progression of the wavefront by Huygens' construction is given by

$$S(x, y, z) = c(x, y, z)t \tag{5.6}$$

the instantaneous positions being given by taking $t = $ constant. In a non-uniform medium, the velocity of the wave v is a function of position, i.e. $v = v(x, y, z)$, where the refractive index $\eta(x, y, z) = c/v$.

The normals to this surface are the rays, and hence

$$\frac{\partial S}{\partial x} = \kappa\alpha \qquad \frac{\partial S}{\partial y} = \kappa\beta \qquad \frac{\partial S}{\partial z} = \kappa\gamma$$

where κ is a factor of proportionality. Hence, moving with the wave

$$c\,dt = dS = \sum_{x,y,z} \frac{\partial S}{\partial x}\,dx = \left(\sum_{x,y,z} \alpha\frac{\partial S}{\partial x}\right)ds = \kappa\,ds \tag{5.7}$$

where ds is an increment of the ray path. Since $ds = v\,dt$ and $\eta = c/v$, then $\kappa = \eta$. Hence

$$\frac{dS}{dx} = \eta\alpha \qquad \frac{dS}{dy} = \eta\beta \qquad \frac{dS}{dz} = \eta\gamma$$

and therefore

$$\left(\frac{dS}{dx}\right)^2 + \left(\frac{dS}{dy}\right)^2 + \left(\frac{dS}{dz}\right)^2 = \eta^2 \tag{5.8}$$

that is, the same relation as is obeyed by the characteristic function V. To prove that these rays are identical with the rays of Fermat's principle, we have

$$\frac{d}{ds}(\eta\alpha) = \frac{d}{ds}\left(\frac{\partial S}{\partial x}\right) = \frac{\partial^2 S}{\partial x^2}\alpha + \frac{\partial^2 S}{\partial y^2}\beta + \frac{\partial^2 S}{\partial z^2}\gamma$$

$$= \frac{1}{\eta}\left(\frac{\partial^2 S}{\partial x^2}\frac{\partial S}{\partial x} + \frac{\partial^2 S}{\partial x\partial y}\frac{\partial S}{\partial y} + \frac{\partial^2 S}{\partial x\partial z}\frac{\partial S}{\partial z}\right)$$

$$= \frac{1}{2\eta} \frac{\partial}{\partial x} \left[\sum_{x,y,z} \left(\frac{\partial S}{\partial x} \right)^2 \right] = \frac{1}{2\eta} \frac{\partial}{\partial x} \eta^2$$

$$= \frac{\partial \eta}{\partial x} \tag{5.9}$$

which follows directly from equation 5.3.

Equation 5.7 shows that the increment in S in passing from one wave surface to an adjacent one is the optical length of the ray; thus S is related to the characteristic function V through

$$S(x, y, z) - S(x', y', z') = V(x, y, z : x', y', z') \tag{5.10}$$

The light rays, being orthogonal to the surfaces of constant S, and by virtue of equations 5.8, will obey the equation

$$\eta \frac{d\mathbf{r}}{ds} = \text{grad } S \tag{5.11}$$

from which it follows that curl $\eta\mathbf{s} = 0$. ($\mathbf{s} = d\mathbf{r}/ds$ is the tangent to the ray; $\eta\mathbf{s}$ is termed the 'ray vector'.)

Hence, by applying Stokes' theorem to any closed curve $\int \eta\mathbf{s}.d\mathbf{l} = 0$, and thus the optical path

$$\int_{P_1}^{P_2} \eta\mathbf{s}.d\mathbf{l}$$

between any two points is independent of the path of integration. This is the 'Lagrange integral invariant'[15] for optical rays. Along a true ray $|d\mathbf{r}| = ds$, and thus the invariant is the optical length given by equation 5.1.

From the ray-tracing point of view, the all-important result is that of equation 5.3. It is therefore of some interest to derive the same result by considering the ray to be the geodesic of a 'refractive space', a concept which will prove to be of value subsequently. We define the refractive space as a space with metric coefficients $G_{\mu\nu} = \eta^2 g_{\mu\nu}$, for $\mu, \nu = 1, 2, 3$, where $\eta(x, y, z)$ is the medium refractive index and $g_{\mu\nu}$ the metric coefficients of the empty space.

The metric is

$$dk^2 = G_{\mu\nu} dx^\mu dx^\nu \tag{5.12}$$

and the geodesics are given by[21]

$$\frac{d^2 x^\mu}{dk^2} + \begin{Bmatrix} \mu \\ \nu \ \lambda \end{Bmatrix} \frac{dx^\nu}{dk} \frac{dx^\lambda}{dk} = 0 \tag{5.13}$$

where the Christoffel symbols are defined by

$$\begin{Bmatrix} \mu \\ \alpha \ \beta \end{Bmatrix} = \tfrac{1}{2} G^{\mu\sigma} \left[\frac{\partial G_{\alpha\sigma}}{\partial x^\beta} + \frac{\partial G_{\sigma\beta}}{\partial x^\alpha} - \frac{\partial G_{\alpha\beta}}{\partial x^\sigma} \right] \tag{5.14}$$

$$\alpha, \beta, \sigma = 1, 2, 3$$

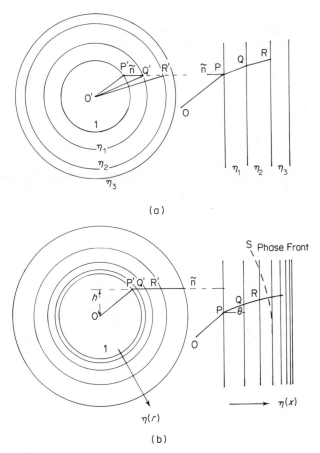

Figure 5.1 (a) Snell's construction for refraction for parallel sided layers of varying refractive index

(b) The construction continued for a continuously variable medium

This definition will form the basis for an inversion transformation which parallels that of Damien's theorem for the single refractive case.

Another curved-ray concept arises if the Snell construction for refraction (Figure 1.9 for example) is extended to the situation where the ray is in a continuously variable medium. We proceed by moving from thin discrete layers to a continuous medium. Thus, as we show in Figure 5.1(a), a parallel-sided system of thin layers has for a corresponding Snell diagram a system of concentric circles. The law of the refractive index in the single direction of variation is $\eta(x)$, and thus the law applied to the circle diagram is $\eta(x) \to \eta(r)$. As each transmitted ray from a given layer is the incident ray for the next, the construction proceeds as follows: O'P' is parallel to the incident ray OP, Q' is obtained as the intersection of (the parallel to) the surface normal ñ with the circle radius $r = \eta_1$, and PQ is parallel to O'Q' (not P'Q'). Similarly QR is parallel to O'R' and the ray PQR is traced by the successive determinations of P'Q'R'. Taking this procedure to the limit of infinitesimally thin layers, as in Figure 5.1(b), we obtain a smooth curve P'Q'R'. In this relation the tangents to the curve PQR are parallel to the radius vectors to the curve P'Q'R'. In the isotropic medium the tangents to the ray are perpendicular to the phase (or wave) front (cf. equations 5.6 and 5.7), the curve S in Figure 5.1(b). There is, therefore, a reciprocity between S and the curve P'Q'R' in which the normals to S are parallel to the radius vectors to P'Q'R' and the normals to P'Q'R' are parallel to the radius vectors to corresponding points on S. In the simple case illustrated here, and for a monotonically varying index of refraction in the axial direction, $\eta = \eta(x)$ transforms into $\eta = \eta(r)$ in the Snell diagram and P'Q'R' is the straight line parallel to the axis. In the medium the ray will be a curve $y = f(x)$, say, and in the Snell diagram P'Q'R' is the line $r = h \operatorname{cosec} \theta$. Then $r = \eta$ and $\tan \theta = f'(x)$.

Consequently

$$f'(x)^2 = \frac{h^2}{\eta^2 - h^2} \qquad \eta = \eta(x)$$

and therefore

$$y = f(x) = \int \frac{h}{(\eta^2 - h^2)^{\frac{1}{2}}} \, dx \tag{5.15}$$

as will be found to be the case in the general expansion of equation 5.3 or by equation 5.13 (see also section 6.1).

Thus the tangents of the ray curve PQR are parallel to the radius vectors OP', OQ', OR' etc. In the isotropic medium these tangents are normals to the wavefront; thus the normals to the wavefront are parallel to the radius vector of the curve in the Snell diagram. If we take the origin of the stratified system to be the point of intersection of the stratified layers, at infinity in the case presented, then the radius vectors to the wavefront at P are parallel to the normals to the curve P'Q'R'. That this is a general situation can be illustrated by making the stratification angular, as shown in Figure 5.2. The rotation of the normals to the stratified layers makes P'Q'R' a curve as shown. As before O'P' is parallel to the tangent at P and thus normal to the phase front at P, and CP is parallel to the normal to the refractive curve at P'.

If the non-uniformity extends to the remaining transverse direction the line P'Q'R' becomes a surface which is called the 'refractive index surface'. The reciprocity we have illustrated between radius vectors and normals of the wavefront and refractive index surface is maintained even when the medium becomes anisotropic,[22] and the ray tangent becomes different from the wavefront-normal.

It can be seen from the simple example that if the tangent at P (Figure 5.1(b)) has direction cosines $\cos \theta$ and $\sin \theta$, the coordinates of P' are $\eta \cos \theta$ and $\eta \sin \theta$. Thus for a wave normal having direction cosines l, m and n, the refractive index surface has direction cosines $(\eta l, \eta m, \eta n)$. This is also the surface of 'normal slowness' and related to the construction of the Fresnel surface.[23] The

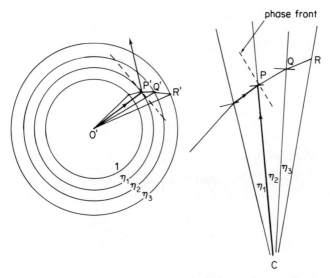

Figure 5.2 Reciprocity in an angularly varying medium

reciprocity then is that proposed by Hamilton based on a previous study by Cauchy.[1(b)]

5.2 EXPANSIONS OF THE RAY EQUATIONS

The equation that has proved to be most applicable to practical problems is equation 5.3, namely

$$\frac{d}{ds}\left(\eta\frac{d\mathbf{r}}{ds}\right) = \nabla\eta$$

and we expand this relation for the coordinate systems in which the ray trajectories will be considered.

5.2.1 Cartesian coordinates

With $\mathbf{r} = x\hat{\mathbf{i}} + y\hat{\mathbf{j}} + z\hat{\mathbf{k}}$

$$\frac{d}{ds}\left[\eta\frac{d}{ds}(x\hat{\mathbf{i}} + y\hat{\mathbf{j}} + z\hat{\mathbf{k}})\right] = \frac{\partial\eta}{\partial x}\hat{\mathbf{i}} + \frac{\partial\eta}{\partial y}\hat{\mathbf{j}} + \frac{\partial\eta}{\partial z}\hat{\mathbf{k}}$$

giving three equations of identical form

$$\frac{d}{ds}\left(\eta\frac{dx}{ds}\right) = \frac{\partial\eta}{\partial x} \tag{5.16}$$

5.2.2 Cylindrical polar coordinates

We require the relations

$$\mathrm{d}\hat{\boldsymbol{\rho}} = -\hat{\boldsymbol{\theta}}\mathrm{d}\theta \qquad \mathrm{d}\hat{\boldsymbol{\theta}} = \hat{\boldsymbol{\rho}}\mathrm{d}\theta$$

Then with $\mathbf{r} = \rho\hat{\boldsymbol{\rho}} + z\hat{\mathbf{k}}$

$$\eta\frac{\mathrm{d}\mathbf{r}}{\mathrm{d}s} = \frac{\mathrm{d}\rho}{\mathrm{d}s}\hat{\boldsymbol{\rho}} - \eta\rho\frac{\mathrm{d}\theta}{\mathrm{d}s}\hat{\boldsymbol{\theta}} + \eta\frac{\mathrm{d}z}{\mathrm{d}s}\hat{\mathbf{k}}$$

and differentiation gives

$$\frac{\mathrm{d}}{\mathrm{d}s}\left(\eta\frac{\mathrm{d}\rho}{\mathrm{d}s}\right) - \eta\rho\left(\frac{\mathrm{d}\theta}{\mathrm{d}s}\right)^2 = \frac{\partial\eta}{\partial\rho} \tag{5.17a}$$

$$\eta\frac{\mathrm{d}\rho}{\mathrm{d}s}\frac{\mathrm{d}\theta}{\mathrm{d}s} + \frac{\mathrm{d}}{\mathrm{d}s}\left(\eta\rho\frac{\mathrm{d}\theta}{\mathrm{d}s}\right) \equiv \frac{1}{\rho}\frac{\mathrm{d}}{\mathrm{d}s}\left(\eta\rho^2\frac{\mathrm{d}\theta}{\mathrm{d}s}\right) = \frac{1}{\rho}\frac{\partial\eta}{\partial\theta} \tag{5.17b}$$

$$\frac{\mathrm{d}}{\mathrm{d}s}\left(\eta\frac{\mathrm{d}z}{\mathrm{d}s}\right) = \frac{\partial\eta}{\partial z} \tag{5.17c}$$

5.2.3 Spherical polar coordinates

We have

$$\mathrm{d}\hat{\mathbf{r}} = \mathrm{d}\theta\hat{\boldsymbol{\theta}} + \sin\theta\mathrm{d}\phi\hat{\boldsymbol{\phi}}$$

$$\mathrm{d}\hat{\boldsymbol{\theta}} = -\mathrm{d}\theta\hat{\mathbf{r}} + \cos\theta\mathrm{d}\phi\hat{\boldsymbol{\theta}}$$

$$\mathrm{d}\hat{\boldsymbol{\phi}} = -\sin\theta\mathrm{d}\phi\hat{\mathbf{r}} - \cos\theta\mathrm{d}\phi\hat{\boldsymbol{\theta}}$$

giving

$$\frac{\mathrm{d}}{\mathrm{d}s}\left(\eta\frac{\mathrm{d}r}{\mathrm{d}s}\right) - \eta r\sin^2\theta\left(\frac{\mathrm{d}\phi}{\mathrm{d}s}\right)^2 - \eta r\left(\frac{\mathrm{d}\theta}{\mathrm{d}s}\right)^2 = \frac{\partial\eta}{\partial r} \tag{5.18a}$$

$$\frac{\mathrm{d}}{\mathrm{d}s}\left(\eta r\frac{\mathrm{d}\theta}{\mathrm{d}s}\right) - \eta r\sin\theta\cos\theta\left(\frac{\mathrm{d}\phi}{\mathrm{d}s}\right)^2 + \eta\frac{\mathrm{d}r}{\mathrm{d}s}\frac{\mathrm{d}\theta}{\mathrm{d}s} = \frac{1}{r}\frac{\partial\eta}{\partial\theta} \tag{5.18b}$$

$$\frac{\mathrm{d}}{\mathrm{d}s}\left(\eta r\sin\theta\frac{\mathrm{d}\phi}{\mathrm{d}s}\right) + \eta r\cos\theta\frac{\mathrm{d}\theta}{\mathrm{d}s}\frac{\mathrm{d}\phi}{\mathrm{d}s} + \eta\sin\theta\frac{\mathrm{d}r}{\mathrm{d}s}\frac{\mathrm{d}\phi}{\mathrm{d}s} = \frac{1}{r\sin\theta}\frac{\partial\eta}{\partial\phi} \tag{5.18c}$$

5.2.4 The general cylindrical medium

We consider a generalized cylindrical system specified by the transformation

$$x = f(u, v) \qquad y = h(u, v) \qquad z = z \tag{5.19}$$

where u, v are orthogonal curvilinear coordinates in a plane perpendicular to the z-axis.

The metric coefficients specified by

$$ds^2 = g_{11}du^2 + g_{22}dv^2 + g_{33}dz^2$$

are therefore

$$g_{11} = f_u^2 + h_u^2 \qquad g_{22} = f_v^2 + h_v^2 \qquad g_{33} = 1$$

where the suffices denote partial differentiation. For orthogonality we have

$$f_u f_v + h_u h_v = 0 \tag{5.20}$$

the product of the Cauchy–Riemann equations. Unit vectors $\hat{e}_u, \hat{e}_v, \hat{e}_z$ are given in terms of basis Cartesian vectors $\hat{e}_x, \hat{e}_y, \hat{e}_z$ by

$$\hat{e}_u = \frac{1}{g_{11}^{\frac{1}{2}}} (f_u \hat{e}_x + h_u \hat{e}_y)$$

$$\hat{e}_v = \frac{1}{g_{22}^{\frac{1}{2}}} (f_v \hat{e}_x + h_v \hat{e}_y); \qquad \hat{e}_z = \hat{e}_z$$

and consequently

$$d\hat{e}_u = \frac{g_{22}^{\frac{1}{2}}}{\Delta g_{11}^{\frac{1}{2}}} [-h_u(f_{uu}du + f_{uv}dv) + f_u(h_{uu}du + h_{uv}dv)]\hat{e}_v$$

$$d\hat{e}_v = \frac{g_{11}^{\frac{1}{2}}}{\Delta g_{22}^{\frac{1}{2}}} [-f_v(h_{vv}dv + h_{uv}du) + h_v(f_{vv}dv + f_{uv}du)]\hat{e}_u$$

where Δ is the Jacobian

$$\Delta = \begin{vmatrix} f_u & h_u \\ f_v & h_v \end{vmatrix}$$

Substituting in the general ray equation for a non-uniform medium with

$$\frac{d\mathbf{r}}{ds} = g_{11}^{\frac{1}{2}} \frac{du}{ds} \hat{e}_u + g_{22}^{\frac{1}{2}} \frac{dv}{ds} \hat{e}_v + g_{33}^{\frac{1}{2}} \frac{dz}{ds} \hat{e}_z \tag{5.22}$$

$$\nabla \eta = \frac{1}{g_{11}^{\frac{1}{2}}} \frac{\partial \eta}{\partial u} \hat{e}_u + \frac{1}{g_{22}^{\frac{1}{2}}} \frac{\partial \eta}{\partial v} \hat{e}_v + \frac{1}{g_{33}^{\frac{1}{2}}} \frac{\partial \eta}{\partial z} \hat{e}_z \tag{5.23}$$

gives the completely general result

$$\frac{d}{ds}\left(\eta g_{11}^{\frac{1}{2}} \frac{du}{ds}\right) + \frac{\eta g_{11}^{\frac{1}{2}}}{\Delta} \frac{dv}{ds}\left[-f_v\left(h_{vv}\frac{dv}{ds} + h_{uv}\frac{du}{ds}\right)\right.$$

$$\left. + h_v\left(f_{vv}\frac{dv}{ds} + f_{uv}\frac{du}{ds}\right)\right] = \frac{1}{g_{11}^{\frac{1}{2}}} \frac{\partial \eta}{\partial u} \tag{5.24a}$$

$$\frac{d}{ds}\left(\eta g_{22}^{\frac{1}{2}} \frac{dv}{ds}\right) + \frac{\eta g_{22}^{\frac{1}{2}}}{\Delta} \frac{du}{ds}\left[-h_u\left(f_{uu}\frac{du}{ds} + f_{uv}\frac{dv}{ds}\right)\right.$$

$$\left. + f_u\left(h_{uu}\frac{du}{ds} + h_{uv}\frac{dv}{ds}\right)\right] = \frac{1}{g_{22}^{\frac{1}{2}}} \frac{\partial \eta}{\partial v} \tag{5.24b}$$

$$\frac{\mathrm{d}}{\mathrm{d}s}\left(\eta g_{33}^{\frac{1}{2}}\frac{\mathrm{d}z}{\mathrm{d}s}\right) = \frac{1}{g_{33}^{\frac{1}{2}}}\frac{\partial\eta}{\partial z} \qquad (5.24c)$$

5.2.5 The general axisymmetric medium

A generalized axisymmetric coordinate system (α, β, ψ) can be specified by the transformation

$$x = F(\alpha, \beta)\cos\psi$$

$$y = F(\alpha, \beta)\sin\psi$$

$$z = H(a, \beta)$$

where ψ is the polar angle of rotation about the z axis and $F(\alpha, \beta)$ and $H(\alpha, \beta)$ are quite general functions of the coordinates (α, β). For the system to be orthogonal we thus require (product of the Cauchy–Riemann equations)

$$F_\alpha F_\beta + H_\alpha H_\beta = 0$$

(suffices denoting partial differentiation). The metric coefficients, defined by

$$\mathrm{d}s^2 = \mathrm{d}x^2 + \mathrm{d}y^2 + \mathrm{d}z^2 = g_{11}\mathrm{d}\alpha^2 + g_{22}\mathrm{d}\beta^2 + g_{33}\mathrm{d}\psi^2$$

are thus

$$g_{11} = F_\alpha^2 + H_\alpha^2 \qquad g_{22} = F_\beta^2 + H_\beta^2 \qquad g_{33} = F^2 \qquad (5.26)$$

The unit vectors $\hat{\mathbf{e}}_\alpha, \hat{\mathbf{e}}_\beta, \hat{\mathbf{e}}_\psi$ are then given in terms of constant Cartesian basis vectors $\hat{\mathbf{e}}_x, \hat{\mathbf{e}}_y, \hat{\mathbf{e}}_z$ by

$$\hat{\mathbf{e}}_\alpha = \frac{(F_\alpha^2 + H_\alpha^2)^{\frac{1}{2}}}{H_\alpha F_\beta - F_\alpha H_\beta}\left[(-H_\beta\cos\psi)\hat{\mathbf{e}}_x - (H_\beta\sin\psi)\hat{\mathbf{e}}_y + F_\beta\hat{\mathbf{e}}_z\right]$$

$$\hat{\mathbf{e}}_\beta = \frac{(F_\beta^2 + H_\beta^2)^{\frac{1}{2}}}{H_\alpha F_\beta - F_\alpha H_\beta}\left[(H_\alpha\cos\psi)\hat{\mathbf{e}}_x + (H_\alpha\sin\psi)\hat{\mathbf{e}}_y - F_\alpha\hat{\mathbf{e}}_z\right]$$

$$\hat{\mathbf{e}}_\psi = -(\sin\psi)\hat{\mathbf{e}}_x + (\cos\psi)\hat{\mathbf{e}}_y$$

The factor $H_\alpha F_\beta - F_\alpha H_\beta$ occurs frequently in the ensuing analysis and we define for brevity

$$H_\alpha F_\beta - F_\alpha H_\beta \equiv K(\alpha, \beta)$$

Differentiation then gives

$$\mathrm{d}\hat{\mathbf{e}}_\alpha = \frac{-K}{H_\alpha(F_\beta^2 + H_\beta^2)^{\frac{1}{2}}}\mathrm{d}\left[\frac{H_\beta(F_\alpha^2 + H_\alpha^2)^{\frac{1}{2}}}{K}\right]\hat{\mathbf{e}}_\beta - \frac{H_\beta(F_\alpha^2 + H_\alpha^2)^{\frac{1}{2}}}{K}\mathrm{d}\psi\hat{\mathbf{e}}_\psi$$

$$\mathrm{d}\hat{\mathbf{e}}_\beta = -\frac{K}{H_\beta(F_\alpha^2 + H_\alpha^2)^{\frac{1}{2}}}\mathrm{d}\left[\frac{H_\alpha(F_\beta^2 + H_\beta^2)^{\frac{1}{2}}}{K}\right]\hat{\mathbf{e}}_\alpha + \frac{H_\alpha(F_\beta^2 + H_\beta^2)^{\frac{1}{2}}}{K}\mathrm{d}\psi\hat{\mathbf{e}}_\psi$$

$$\mathrm{d}\hat{\mathbf{e}}_\psi = \left[-\frac{F_\alpha}{(F_\alpha^2 + H_\alpha^2)^{\frac{1}{2}}}\hat{\mathbf{e}}_\alpha - \frac{F_\beta}{(F_\beta^2 + H_\beta^2)^{\frac{1}{2}}}\hat{\mathbf{e}}_\beta\right]\mathrm{d}\psi \qquad (5.27)$$

Hence from equation 5.22

$$\frac{d\mathbf{r}}{ds} = (F_\alpha^2 + H_\alpha^2)^{\frac{1}{2}}\frac{d\alpha}{ds}\,\hat{\mathbf{e}}_\alpha + (F_\beta^2 + H_\beta^2)^{\frac{1}{2}}\frac{d\beta}{ds}\,\hat{\mathbf{e}}_\beta + F\frac{d\psi}{ds}\,\hat{\mathbf{e}}_\psi \qquad (5.28)$$

and from equation 5.23

$$\nabla\eta = \frac{1}{(F_\alpha^2 + H_\alpha^2)^{\frac{1}{2}}}\frac{\partial\eta}{\partial\alpha}\,\hat{\mathbf{e}}_\alpha + \frac{1}{(F_\beta^2 + H_\beta^2)^{\frac{1}{2}}}\frac{\partial\eta}{\partial\beta}\,\hat{\mathbf{e}}_\beta + \frac{1}{F}\frac{\partial\eta}{\partial\psi}\,\hat{\mathbf{e}}_\psi \qquad (5.29)$$

Differentiation of equation 5.28 gives the final general result

$$\frac{d}{ds}\left[\eta(F_\alpha^2 + H_\alpha^2)^{\frac{1}{2}}\frac{d\alpha}{ds}\right] - \eta\,\frac{K(F_\beta^2 + H_\beta^2)^{\frac{1}{2}}}{H_\beta(F_\alpha^2 + H_\alpha^2)^{\frac{1}{2}}}\frac{d}{ds}\left[\frac{H_\alpha(F_\beta^2 + H_\beta^2)^{\frac{1}{2}}}{K}\right]\frac{d\beta}{ds}$$

$$-\eta\,\frac{FF_\alpha}{(F_\alpha^2 + H_\alpha^2)^{\frac{1}{2}}}\left(\frac{d\psi}{ds}\right)^2 = \frac{1}{(F_\alpha^2 + H_\alpha^2)^{\frac{1}{2}}}\frac{\partial\eta}{\partial\alpha} \qquad (5.30a)$$

$$\frac{d}{ds}\left[\eta(F_\beta^2 + H_\beta^2)^{\frac{1}{2}}\frac{d\beta}{ds}\right] - \eta\,\frac{K(F_\alpha^2 + H_\alpha^2)^{\frac{1}{2}}}{H_\alpha(F_\beta^2 + H_\beta^2)^{\frac{1}{2}}}\frac{d}{ds}\left[\frac{H_\beta(F_\alpha^2 + H_\alpha^2)^{\frac{1}{2}}}{K}\right]\frac{d\alpha}{ds}$$

$$-\eta\,\frac{FF_\beta}{(F_\beta^2 + H_\beta^2)^{\frac{1}{2}}}\left(\frac{d\psi}{ds}\right)^2 = \frac{1}{(F_\beta^2 + H_\beta^2)^{\frac{1}{2}}}\frac{\partial\eta}{\partial\beta} \qquad (5.30b)$$

$$\frac{d}{ds}\left(\eta F\frac{d\psi}{ds}\right) - \eta\,\frac{H_\beta(F_\alpha^2 + H_\alpha^2)}{K}\frac{d\alpha}{ds}\frac{d\psi}{ds} + \eta\,\frac{H_\alpha(F_\beta^2 + H_\beta^2)}{K}\frac{d\beta}{ds}\frac{d\psi}{ds} = \frac{1}{F}\frac{\partial\eta}{\partial\psi}$$

$$(5.30c)$$

From the orthogonality condition and the definition of K we find also that

$$F_\alpha^2 + H_\alpha^2 = \frac{H_\alpha K}{F_\beta} \qquad F_\beta^2 + H_\beta^2 = \frac{F_\beta K}{H_\alpha}$$

With circular symmetry, $\partial\eta/\partial\psi = 0$ and equation 5.30c reduces to

$$\frac{d}{ds}\left(\eta F\frac{d\psi}{ds}\right) + \eta F_\alpha\frac{d\alpha}{ds}\frac{d\psi}{ds} + \eta F_\beta\frac{d\beta}{ds}\frac{d\psi}{ds} = 0 \qquad (5.31)$$

or

$$\frac{1}{F}\frac{d}{ds}\left(\eta F^2\frac{d\psi}{ds}\right) = 0$$

$\eta F^2(d\psi/ds) = \kappa$ is a ray constant called the 'skew invariant' of the ray.

6 Rays in Linear and Cylindrical Media

We can now deal with those solutions of the ray equations which give rise to practical focusing devices or systems in which we need to derive the ray trajectory within a given medium. In the main our attention will be concentrated on the cylindrical and spherical coordinate systems, and on refractive indices in a realistic range, that is real and greater than unity (less than a value only a few times this). In microwave practice this range can be extended somewhat, although the densities required and the bulk of material rapidly becomes excessive in all but the very short wavelength range.

However, for purely theoretical purposes it is of great interest to consider the entire range of (real) values of refractive index, including values less than unity, zero and infinite. Values less than unity occur in practice in the microwave waveguide analogue of propagation of a non-T.E.M. mode between parallel conducting plates.[24] With such a range of values we can retain the concept of the 'refractive space' which has particular ray properties which are subject to transformations and subsequently from which 'real' regions can be selected to provide practical focusing devices. By retaining the complete refractive space, other theoretical considerations lead to strong interconnections with analogous problems in the other least-action subject, namely, particle trajectories in non-uniform potentials.

It is found, in general, that solutions of the ray equations are most easy to derive when the ray is considered to be confined to a surface for which one coordinate of the system has a constant value—that is, on meridional planes, planes containing the axis of symmetry in a cylindrical system or, of course, cylinders coaxial with the refractive system. The spherical coordinate system has the greatest ramifications in the theoretical process and will thus be considered in greater detail in Chapter 7.

Some general laws can be enunciated at the outset. A ray can, and often does, at some point of its trajectory, become parallel to the general stratification of the medium. A ray can only achieve a direction perpendicular to the stratification where the refractive index is infinite. A ray which is perpendicular to the stratification at its source will thus remain so for its total trajectory. At an interface between two non-uniform media, the ray will obey Snell's law of refraction at each point, as though the media were infinite and with values obtaining at that point, the so-called 'geometrical optics' approximation. Reflections of rays at such surfaces, as is the common practice in optics, are ignored.

6.1 THE LINEAR STRATIFIED MEDIUM

We consider first the stratification to depend upon the single coordinate x and the ray to lie in the (x, y) plane. This reproduces the conditions of Figure 5.1. Then $ds^2 = dx^2 + dy^2$ and equations 5.16 become, since $\eta = \eta(x)$

$$\frac{d}{ds}\left(\eta \frac{dx}{ds}\right) = \frac{d\eta}{dx}$$

$$\frac{d}{ds}\left(\eta \frac{dy}{ds}\right) = 0 \tag{6.1}$$

Both equations lead to the same result, on substituting $d/ds = \cos \psi(d/dx)$, namely

$$\eta \frac{dy}{ds} = \text{constant} = h \text{ along a ray}$$

Since $dy/ds = \sin \psi$ (Figure 6.1) this is simply the continuation of Snell's law, $\eta \sin \psi = h$. Then

$$dy = \frac{h dx}{\{\eta^2 - h^2\}^{\frac{1}{2}}} \quad \text{or} \quad y = \int \frac{h dx}{\{\eta^2 - h^2\}^{\frac{1}{2}}} \tag{6.2}$$

as was shown by equation 5.15.

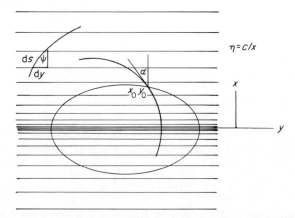

Figure 6.1 Rays in a medium with refractive index varying with the law $\eta = c/x$

This ray trajectory can be treated on two levels, one given the refractive index to determine the ray path. For example, consider the law $\eta(x) = c/x$ and a ray originating at the point $x = x_0$, $y = 0$; at which point the refractive index is $\eta_0 = c/x_0$, making an angle α with the vertical. Hence the ray constant h is $\eta_0 \sin \alpha = c \sin \alpha/x_0$. Substituting in equation 6.2 and integrating we obtain the trajectories

$$(y\eta_0^2 \sin^2 \alpha + b)^2 = c^2 - \eta_0 x^2 \sin^2 \alpha$$

b being an arbitrary constant of integration. With $b = c$ these are the equations of ellipses in general, and for the particular ray for which $c \sin \alpha / x_0 = 1$ the trajectory is a circle. The ellipses are all such that one axis lies along the line $x = 0$, that is they cross the infinite refractive index value perpendicularly (Figure 6.1).

The alternative is to specify the ray trajectory and to derive the refractive index law. For this, for example, we state *a priori* that all the rays from a source on the axis $x = 0$ should refocus at the point $y = 2a$. By symmetry the ray must then be parallel to the stratification at the point $y = a$ (Figure 6.2). We therefore have in equation 6.1

$$a = \int \frac{h\mathrm{d}x}{(\eta^2 - h^2)^{\frac{1}{2}}} \qquad h = \eta_0 \sin \alpha$$

where η_0 is the value of η at $x = 0$.

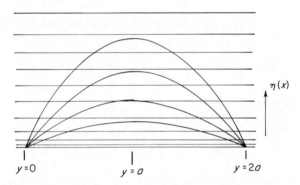

Figure 6.2 Rays in the medium with refractive-index law $\eta = \eta_1 \operatorname{sech} (\pi x/2a)$ refocus periodically along the axis

The method given in Appendix III for the conversion of these equations to Abel's equation, and subsequent solution, results in the law

$$\eta(x) = \eta_0 \operatorname{sech} \frac{\pi x}{2a} \tag{6.3}$$

and substitution of this result into equation 6.1 and integration gives the trajectory

$$\sin \frac{\pi y}{2a} = \tan \alpha \sinh \frac{\pi x}{2a} \tag{6.4}$$

with constant of integration appropriate to a source at $x = 0$.

Luneburg,[25] whose work in this field predominates, shows that an off-axis displacement of the source does not affect the refocusing of the rays, of necessity, to an image off-axis position as shown in Figure 6.3, since substitution of the law of equation 6.3 into the differential form of 6.2 gives

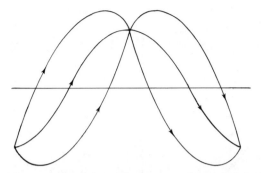

Figure 6.3 Refocusing from an offset source in the sech-law medium

$$dy = \frac{\sin \alpha \cosh\left(\dfrac{\pi x}{2a}\right)dx}{\left[\cos^2 \alpha - \sin^2 \alpha \sinh^2\left(\dfrac{\pi x}{2a}\right)\right]^{\frac{1}{2}}} \tag{6.5}$$

which, being independent of η_0, is independent of the x coordinate of the source x_0.

6.2 THE CYLINDRICAL NON-UNIFORM MEDIUM

6.2.1 The circularly symmetrical medium

The most commonly found of all non-uniform materials are the glass fibres currently used for optical transmission. For a rotationally symmetrical medium uniform in the longitudinal z direction, we put in equations 5.17

$$\frac{\partial \eta}{\partial \phi} = 0 \qquad \frac{\partial \eta}{\partial z} = 0$$

We obtain $(d/ds)[\eta(dz/ds)] = 0$ identical to equation 6.1, and from 5.17b

$$\frac{d}{ds}\left(\eta\rho^2 \frac{d\theta}{ds}\right) = 0 \tag{6.6}$$

Thus if the rays are confined to meridional planes and $d\theta/ds = 0$ as a consequence, the system is the equivalent of the rotation about the axis $x = 0$ of the linear system in section 6.1. Thus the cylindrical form with refractive index

$$\eta(\rho) = \eta_0 \operatorname{sech}\left(\frac{\pi\rho}{2a}\right) \tag{6.7}$$

will have the same trajectories, in the meridional planes only, shown in Figure 6.2.

At any position midway between two perfect foci, the rays by symmetry alone will be parallel to the stratification; thus if the medium were to be terminated

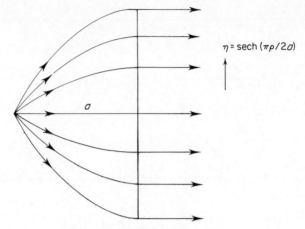

Figure 6.4 The short-focus horn

there with a place surface perpendicular to the axis, the rays would continue into free space as a parallel beam. This design of lens was first proposed by Brown[26] and termed the 'short-focus horn' (Figure 6.4).

The medium could, of course, be terminated by any other shaped surface to provide a local perfect focus, or indeed any specified caustic, symmetric or asymmetric.

If the rays are no longer confined to the meridional plane, equation 6.6 gives a second constant, applicable to the now skew ray. That is

$$\eta \rho^2 \frac{d\theta}{ds} = \text{constant} = h \tag{6.8}$$

the Herzberger 'skew invariant'.[27]

The implication is that when h is zero the rays are once again confined to meridional planes, and thus equation 6.8 contains a measure of the skew behaviour of these rays. This is borne out by the analysis given by Luneburg. Recasting equation 6.8 as a dependent function of the axial variable z, i.e. substituting $ds^2 = d\rho^2 + \rho^2 d\phi^2 + dz^2$, one obtains Luneburg's equation (30.15)

$$\frac{\eta \rho^2 \dfrac{d\phi}{dz}}{\left[1 + \left(\dfrac{d\rho}{dz}\right)^2 + \rho^2 \left(\dfrac{d\phi}{dz}\right)^2\right]^{\frac{1}{2}}} = h$$

Solving this for $d\phi/dz$ and integrating gives

$$\phi_1 - \phi_0 = h \int_{z_1}^{z_0} \frac{\left[1 + (d\rho/dz)^2\right]^{\frac{1}{2}}}{\rho(\eta^2 \rho^2 - h^2)^{\frac{1}{2}}} \, dz \tag{6.9}$$

where (ϕ_0, z_0) and (ϕ_1, z_1) are the *angular* positions at the end-points of a ray.

We now separate the optical path of a ray from a point in the z_0 plane to a point in the z_1 plane, into an equivalent 'radial' path and an 'angular' optical path, the latter given by $h(\mathrm{d}\phi/\mathrm{d}z)$. Then the 'radial' path will be given by the total path minus the 'angular' path, that is

$$\int_{z_0}^{z_1} \left\{ \eta \left[1 + \left(\frac{\mathrm{d}\rho}{\mathrm{d}z}\right)^2 + \rho^2 \left(\frac{\mathrm{d}\phi}{\mathrm{d}z}\right)^2 \right]^{\frac{1}{2}} - h \left(\frac{\mathrm{d}\phi}{\mathrm{d}z}\right) \right\} \mathrm{d}z \qquad (6.10)$$

which equals, after substituting for $\mathrm{d}\phi/\mathrm{d}z$

$$\int_{z_0}^{z_1} \left(\eta^2 - \frac{h^2}{\rho^2}\right)^{\frac{1}{2}} \left[1 + \left(\frac{\mathrm{d}\rho}{\mathrm{d}z}\right) \right]^{\frac{1}{2}} \mathrm{d}z \qquad (6.11)$$

which is (ρ, z) dependent only; hence the definition of 'radial' optical path.

This is equivalent to the path of a ray in a *meridional* plane in a medium with refractive index

$$\mu(\rho, z) = \left(\eta^2 - \frac{h^2}{\rho^2}\right)^{\frac{1}{2}} \qquad (6.12)$$

Thus for a skew ray we obtain the radial movement, with respect to distance along the axis, from equation 6.11 as for a Cartesian geometry but using the refractive index $\mu(\rho, z)$, and the angular motion from equation 6.12.

6.2.2 Flat-disc lenses

A similar collimating lens can be obtained in a considerably reduced body of material if the rays are allowed their natural spread in free space up to some plane surface A, as shown in Figure 6.5, and curve thereafter to become horizontal at a

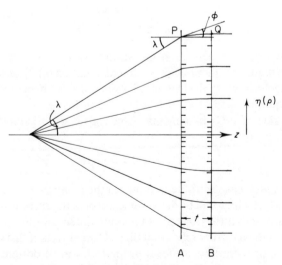

Figure 6.5 Non-uniform flat-disc lens

second plane surface B, assisted by the refraction that occurs at the surface A. This creates a flat-disc radially non-homogeneous lens with an external source.

A lens of this type first described by Wood[28] made use of a paraxial approximation and the assumption that, within the body of the lens, the rays were arcs of circles. As will be seen, this assumption gives a very good indication of the form of refractive index law required. The result here is for a wide-angle lens $f/D \simeq 1$, and consequently for exact ray paths.

Once established the method is adaptable to lenses with other focal requirements and to multi-lens systems.

We consider, as in Figure 6.5, a ray from a source on the axis at *unit* focal distance OP from the plane surface A perpendicular to the axis. This ray makes an angle λ with the axis, and at P undergoes a refraction in accordance with Snell's law. We assume this refraction to occur as though the medium at P were infinite, and of uniform refractive index of value $\eta(P)$. This is the usual assumption for non-uniform surfaces or media in physical optics.

At the point P, therefore, we have

$$\eta(P) \sin \phi = \sin \lambda$$

and hence

$$\sin \phi = \frac{\rho}{(1 + \rho^2)^{\frac{1}{2}} \eta} \qquad (6.13)$$

At P, $\eta \cos \phi = A$, and therefore the ray constant for the ray OPQ is

$$A = \left(\eta^2 - \frac{\rho^2}{1 + \rho^2} \right)^{\frac{1}{2}} \quad \text{evaluated at P} \qquad (6.14)$$

This function will be designated $A(P)$. Within the medium itself we use equation 6.2

$$z = \int_{\rho(P)}^{\rho(Q)} \frac{A(P)\mathrm{d}\rho}{\left[\eta^2(\rho) - A^2(P)\right]^{\frac{1}{2}}} \qquad (6.15)$$

In order to obtain a second plane surface where the rays have become horizontal, we require a function $\eta(\rho)$ such that equation 6.15 results in a constant value $z = t$, the thickness of the lens, for all values of λ and hence for all values of ϕ as given by equation 6.13.

Since the value of $\rho(Q)$ is then that which gives a horizontal ray point, or $\mathrm{d}z/\mathrm{d}s \equiv 1$, then at Q

$$A(Q) \equiv A(P) = \eta(Q)$$

In other words $\rho(Q)$ has the value that makes the denominator in equation 6.15 equal to zero, a fact which will be made some use of in the ensuing analysis. It is, of course, understood that the objective in making the second surface B plane, is to not have to consider a further refraction there. Other lenses of this kind, with non-planar surfaces, or with different focal properties, can be designed by a similar analysis extended by including a refraction at the second surface.

The solutions of equation 6.15 for the complete medium, that is the short-focus horn, has $\rho(P)$ zero and η can be obtained by the standard method of converting equation 6.15 to an integral of Abel's form (Appendix III). With the integral now incomplete these methods no longer apply and the approach used is to insert such trial functions of $\eta(\rho)$ as makes equation 6.15 integrable by analytic methods. It is soon found that quadratic and quartic functions of both $\eta^2(\rho)$ and $1/\eta^2(\rho)$ can be dealt with by the use of incomplete elliptic integrals of the first kind.[29]

The closest approximation to a plane-surfaced lens has come from modifications to the parabolic law of refractive index $\eta = a - b\rho^2$. Besides being the most commonly used law in gradient index fibres, it is the result obtained by Wood himself in the approximate solution previously mentioned.

This can be extended by applying a law of the form

$$\eta^2(\rho) = a^2 - 2b^2\rho^2 + c^2\rho^4$$

Then the constant lens thickness is obtained from

$$t = \int_{\rho(P)}^{\rho(Q)} \frac{A d\rho}{[(a^2 - A^2) - 2b^2\rho^2 + c^2\rho^4]^{\frac{1}{2}}}$$

$$= \int_{\rho(P)}^{\rho(Q)} \frac{A d\rho}{c[(\alpha^2 - \rho^2)(\beta^2 - \rho^2)]^{\frac{1}{2}}} = \frac{A}{c} I \qquad (6.16)$$

where

$$\alpha^2 = \{b^2 + [b^4 - c^2(a^2 - A^2)]^{\frac{1}{2}}\}/c^2$$

$$\beta^2 = \{b^2 - [b^4 - c^2(a^2 - A^2)]^{\frac{1}{2}}\}/c^2$$

and from equation 6.14

$$A^2 = a^2 - 2b^2\rho^2 + c^2\rho^4 - \rho^2/(1 + \rho^2)| \text{ at } \rho = \rho(P)$$

Then by the virtue of the fact that $\rho(Q)$ is a root of the denominator, it must equal β, and hence with

$$\alpha > \beta > \rho(P) > 0$$

equation 6.16 fulfils the conditions for the incomplete elliptic integral of the first kind,[29] and

$$I = gF(\psi, k)$$

where

$$g = \frac{1}{\alpha}; \quad \psi = \sin^{-1} \left\{ \frac{\alpha^2[\beta^2 - \rho^2(P)]^{\frac{1}{2}}}{\beta^2[\alpha^2 - \rho^2(P)]^{\frac{1}{2}}} \right\} \qquad (6.17)$$

and $k^2 = \beta^2/\alpha^2$. In the limit $\rho \to 0$, $a \to A$ and $k \to 0$, and therefore

$$\psi \to \psi_0 = \sin^{-1} \left(\frac{1}{1 + 2b^2} \right)^{\frac{1}{2}}$$

This gives the axial thickness of the lens

$$t_0 = \frac{a}{\sqrt{(2b^2)}} F(\psi_0, 0)$$

Since, with $k = 0$, $F(\psi_0, 0) \simeq \sin \psi_0$ over the range usually encountered

$$t_0 \simeq \frac{a}{(2b^2 + 4b^4)^{\frac{1}{2}}}$$

The requirement is therefore for t of equation 6.16 to be made constant in value for all values of $\rho(P)$ and equal to t_0, by adjusting the arbitrary constants a, b and c. This can be achieved by optimization processes such as minimizing the value of

$$(t^2 - t_0^2)^{\frac{1}{2}}$$

Subsequent ray-tracing computation shows that for $\eta^2 = 16 - 24\rho^2 + 12.85\rho^4$, the rays are horizontal (to $1:10^{-6}$) over a surface which is flat to within 0.0004. (OF $= 1$) over the range $0 < \rho < 0.5$, that is for an f/D ratio of unity. The thickness is approximately $f/6$.

The high degree of symmetry arising from the plane surfaces of the flat-disc lens opens up many opportunities. The lens focuses identically from sources on either side, for example, and the insertion of a mid-plane reflector converts it into a corrected plane reflector (Figure 6.6). Owing to the simplicity with which

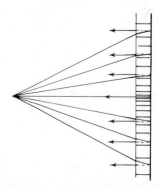

Figure 6.6 The bisected disc lens as a coated plane reflector

refraction can be dealt with at plane surfaces, multi-lens systems are conceivable for such effects as variable beam shaping, zoom focusing and aberration correction. Other applications could include focusing end-plates for laser cavities, matching junctions for gradient index fibres and large reflector feeding systems. The method is, of course, particularly applicable to acoustic lenses with similar focusing requirements.

6.2.3 Flat-lens doublets

One of the major advantages of the flat-disc radially non-homogeneous lens is that the range of incidence angles at the source is reduced for comparable f/D

ratios. A penalty for this is the loss of an important refractive-index scaling property that applies to cylindrical systems where the source is embedded in the variable medium. This property can be recovered, and the range of incidence angles at the source can be further reduced by the use of a lens doublet, consisting of two parallel-sided discs in contact. The two-layer cylindrical lens then has to consist of cylindrically symmetrical non-homogeneous media, in the first layer of which the rays are given an *upward* curvature by the appropriate choice of the law of the refractive index, and in the second layer a refractive-index law which returns the rays to the axial direction assisted, usually, by a refraction at the interface between the two.

The upward curvature of the rays in the first region can be obtained by two methods: first by a radially non-homogeneous medium in which the refractive index increases with the radius or, alternatively, by a radially homogeneous medium with refractive index decreasing in the axial direction. The analyses for both systems have much in common, and utilize the ray-trace integral for the appropriate refractive-index variation. The radial refractive-index variation in the second region has then to be a decreasing function of the radius, and is chosen so that the rays all become axial at points which lie on a plane perpendicular to the axis. It is shown that this condition can be met to within a small tolerance by the use of analytic functions for the refractive index.

The double radial lens is illustrated in Figure 6.7. The first region from the source S to the plane surface F has a refractive-index law $\eta_1(\rho)$ which is an increasing function of ρ causing the rays to curve upwards as shown. At P, on the interface between the two regions, the ray is refracted. In the second region the refractive-index law $\eta_2(\rho)$ is a decreasing function of ρ. The rays continue to curve towards the axis, eventually becoming parallel to it. The design procedure is to arrange the two separate refractive-index variations so that the rays become parallel to the axis at points which lie on a plane surface parallel to F.

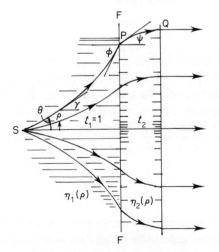

Figure 6.7 The flat-lens doublet

The ray constant for a cylindrically varying non-homogeneous medium is

$$\eta(\rho) \cos \gamma = A$$

where γ is the angle between the tangent to the ray and the axis at any point on the ray with radial coordinate ρ, as shown. At P, therefore

$$\eta_1(P) \cos \phi = A$$

and, for refraction there, we have

$$\eta_1(P) \sin \phi = \eta_2(P) \sin \psi$$

The ray constant B for the path from P to Q in the second medium is, therefore, given by

$$B = \eta_2(P) \cos \psi$$

which becomes

$$B = [\eta_2^2(P) - \eta_1^2(P) + A^2]^{\frac{1}{2}}$$

The thicknesses of the two media are given by the two integrals of the ray equation

$$t_1 = \int_S^P \frac{A\,\mathrm{d}\rho}{(\eta_1^2 - A^2)^{\frac{1}{2}}} \tag{6.18a}$$

$$t_2 = \int_P^Q \frac{B\,\mathrm{d}\rho}{(\eta_2^2 - B^2)^{\frac{1}{2}}} \tag{6.18b}$$

We can normalize the thickness of the first medium to unity making $t_1 = 1$; all dimensions will be referred to this value hereafter.

The ray becomes horizontal again, i.e. $\gamma = 0$, in the second medium where $\eta_2(Q) = B$, that is, where the denominator in the integrand in equation 6.18b has a zero value. That is, Q is a solution of

$$\eta_2(\rho) = B$$

We can now substitute arbitrary, but integrable, functions of $\eta_1(\rho)$ into equation 6.18a, with ray 'starting' conditions at S determining the value of A; and thus obtain the value of P. The constant B for equation 6.18b will then be defined for selected functions $\eta_2(\rho)$ and the thickness t_2 derived. The refractive law $\eta_1(\rho)$ has to be an increasing function of ρ, integrable in the form of equation 6.18a. In the following we utilize the simple quadratic form

$$\eta_1^2(\rho) = a^2 + b^2\rho^2$$

although other forms such as

$$\eta^2(\rho) = a^2/(1 - b\rho)^2$$

could also be used. In equation 6.18b quadratic forms also give results in terms of elementary integrals, but can readily be shown always to result in values of t_2 which are decreasing functions and thus always give a tapered second surface.

The quartic form

$$\eta_2^2(\rho) = p^2 - c^2\rho^2 + d^2\rho^4 \qquad (6.19)$$

allows the integration of equation 6.18b in terms of the incomplete elliptic integrals and gives results, shown later, with a high degree of flatness for the final surface.

A double-layer lens of this type is illustrated in Figure 6.8, for which the input rays are all horizontal over a small (with respect to the unit dimension t_1) circular spot about the axis. This can be magnified by the radial construction into a much

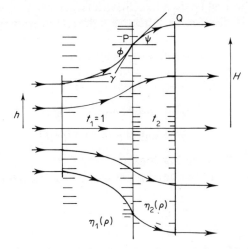

Figure 6.8 The flat-lens doublet as a telescopic lens

larger circular aperture with horizontal rays, a non-uniform telescopic lens. In reverse the lens concentrates normally incident rays over the large aperture onto the smaller spot, retaining normal rays at the focus.

The ray constant is dependent only on the height of the ray, which, being parallel to the axis, is thus

$$A = \eta_1(h)$$

Then at P, for a unit thickness of the first region

$$1 = \int_h^P \frac{\eta_1(h)\,d\rho}{[\eta_1^2(\rho) - \eta_1^2(h)]^{\frac{1}{2}}}$$

Choosing $\eta_1^2(\rho)$ to be $a^2 + b^2\rho^2$ we obtain

$$P = h \cosh\left[b/(a^2 + b^2h^2)^{\frac{1}{2}}\right]$$

At P, $\eta_1(P)\sin\phi = \eta_2(P)\sin\psi$, where $\eta_1(P)\cos\phi = A = \eta_1(h)$. Then the ray constant for the second medium is

$$B = \eta_2(P)\cos\psi = \{\eta_2^2(P) - b^2h^2\sinh^2\left[b/(a^2 + b^2h^2)^{\frac{1}{2}}\right]\}^{\frac{1}{2}}$$

For the second medium we substitute

$$\eta_2(\rho) = a^2 + b^2 - c^2\rho^2 + d^2\rho^4$$

and the procedure is then the same as for the single lens.

The degree of ray bending is due to the variable terms ρ^2 and ρ^4, and the constants can therefore be chosen so that the refractive index is always lower in the second medium than in the first. This is not essential to the analysis, but is logical from physical considerations, and will give a thinner overall lens.

The 'crossover' can be arranged at an arbitrary value of the radius, say at $\rho = 1$. This choice implies $d^2 = c^2$, for example. A refraction away from the axis is still permissible and could well be used to achieve a larger exit aperture and a lower f/D ratio. Too rapid decrease in refractive index could lead to total internal reflection. This would show up as a breakdown in the computational procedure, through the occurrence of a negative value for B^2.

If we define

$$\begin{aligned}
D^2 = &\, c^2 h^2 \cosh^2 \left[b/(a^2 + b^2 h^2)^{\frac{1}{2}}\right] \\
&- d^2 h^4 \cosh^4 \left[b/(a^2 + b^2 h^2)^{\frac{1}{2}}\right] \\
&+ b^2 h^2 \sinh^2 \left[b/(a^2 + b^2 h^2)^{\frac{1}{2}}\right]
\end{aligned}$$

then $B^2 = a^2 + b^2 - D^2$ is the ray constant.

The thickness up to the horizontal point of any ray is given by (cf. equation 6.16)

$$\begin{aligned}
t_2 &= \int_P^Q \frac{(a^2 + b^2 - D^2)^{\frac{1}{2}}\mathrm{d}\rho}{(d^2\rho^4 - c^2\rho^2 + D^2)^{\frac{1}{2}}} \\
&\equiv \int_P^Q \frac{(a^2 + b^2 - D^2)^{\frac{1}{2}}\mathrm{d}\rho}{d[(\alpha^2 - \beta^2)(\beta^2 - \rho^2)]^{\frac{1}{2}}}
\end{aligned}$$

where

$$\alpha^2 = \left[c^2 + (c^4 - 4D^2 d^2)^{\frac{1}{2}}\right]/2d^2$$

$$\beta^2 = \left[c^2 - (c^4 - 4D^2 d^2)^{\frac{1}{2}}\right]/2d^2$$

Then since $\rho(Q)$ has to be a zero value of the denominator at the point the ray is horizontal

$$\rho(Q) \equiv \beta$$

Then

$$t_2 = [a^2 + b^2 - D^2]^{\frac{1}{2}} F(\psi, k)/\alpha d$$

where $F(\psi, k)$ is the incomplete elliptic integral of the first kind

$$\psi = \sin^{-1}\left[\frac{\alpha^2(\beta^2 - P^2)}{\beta^2(\alpha^2 - P^2)}\right]^{\frac{1}{2}} \qquad k^2 = \beta^2/\alpha^2 \qquad (6.20)$$

The radius of the final aperture is H, the maximum value of β obtained over which a specified degree of flatness, that is constancy of t_2, is realized, for a given input spot radius $h_{\max}$.

The partial degree of freedom lost, in taking p^2 in equation 6.19 to be dependent on the constants of the first medium only, alters the overall dimension of the lens and not the flatness of the final surface. It can be seen that, with the choice of refractive-index laws made, all the above equations contain ratios only of the parameters. Hence, once a solution for a flat surface is obtained, the results can be scaled to high or low values of refractive index without altering the dimensions arrived at in the first instance. This would occur for any solution derived from refractive-index laws with homogeneous powers of the parameters. The best result achieved, using unsophisticated computational methods, shows a deviation of ± 0.0002 for an output radius of 0.5, for a radial magnification of 2.5:1. As expected, it is possible to increase this radius at the expense of the maximum deviation from flatness. By virtue of the exact ray path analysis, the final surface is itself an equiphase surface, and thus the phase error follows the same profile, which for the results given is shown in Figure 6.9.

It is also possible to use a second degree of freedom by allowing the interface to be curved as shown in Figure 6.10. If this curve is a specified function $f(\rho, z) = 0$, a solution for $\rho(P)$ will still be possible for a given refractive-index law. Hence the continuation to a final plane surface will take the same form as in the analysis given. It should be noted that this is not the case for a specified curved final surface, the analysis being heavily dependent on the horizontality of the ray at the output surface. The additional degree of freedom will allow for the inclusion of such effects as agreement with the sine law or specification of amplitude distributions.

Similar results occur when an isotropic source is placed on the axis, but the ray

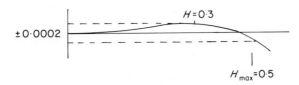

Figure 6.9 Residual phase error in the flat-lens telescopic doublet

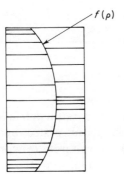

Figure 6.10 Shaping of the intermediate surface of the flat-lens doublet

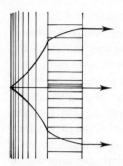

Figure 6.11 Flat-lens axial–radial doublet. Rays in the first region acquire their upward curvature from the decrease in refractive index in the axial direction

solution is confined to a small angle about the axis. This would normally be termed a paraxial solution, but we are nonetheless concerned with lenses with f/D ratios between 0.5 and unity. Two types of lenses are possible, a doubly radial type, as in the previous example, and a type termed the axial–radial doublet (Figure 6.11). In the latter the upward curvature of the rays is produced by a refractive-index law which *decreases* along the axial direction.

Only the conditions governing the derivation of the ray constant B for the second medium differ. For the radial medium

$$\eta_1^2(\rho) = a^2 + b^2\rho^2$$

the source condition for a ray making an angle θ (Figure 6.7) with the axis is

$$A = \eta_1(0)\cos\theta = a\cos\theta$$

Owing to refraction at P the ray constant for the second medium is

$$B^2 = \eta_2^2(P) - a^2\sin^2\theta\cosh^2\left(\frac{b}{a\cos\theta}\right)$$

With the refractive index z dependent in the first medium, we require the *decreasing* function (Figure 6.11)

$$\eta_1^2(z) = a^2 - b^2 z^2$$

The ray constant is then $A = \eta_1(0)\sin\theta = a\sin\theta$ (note) and P derives from, with unity first thickness

$$P = \int_0^1 \frac{A\,dz}{[\eta_1^2(z) - A^2]^{\frac{1}{2}}} = \frac{a}{b}\sin\theta\sin^{-1}\left(\frac{b}{a\cos\theta}\right)$$

Refraction at P gives

$$B^2 = \eta_2^2(P) - A^2 = \eta_2^2(P) - a^2\sin^2\theta$$

and the method proceeds as before with these values inserted.

If a transverse section only of such lenses is taken, a strip construction follows suitable for the same beam-forming mechanism in an integrated optical component. All the focusing properties available in the complete cylindrical system then become possible in the non-uniform layer over a substrate that is the foundation of the integrated optical system. The scaling property of the refractive index allows low values to be used without altering the dimensions of the component.

6.2.4 The general cylindrical medium

The coordinate system for the general cylinder system is given in Section 5.2.4, where $f(u, v)$ and $h(u, v)$ are generalized orthogonal coordinates in a plane. Such coordinate systems can be derived directly from the conformal mapping

$$z \equiv x + \mathrm{i}y = F(u + \mathrm{i}v)$$

$$= f(u, v) + \mathrm{i}h(u, v) \equiv F(w) \tag{6.21}$$

although conformality in addition to orthogonality, equation 5.20, is not essential and would prove an unnecessary restraint on the selection of cross-sectional geometries. This can be seen by comparing transformations 1 and 3 in the Table, where the first is orthogonal but not conformal whereas the third is both.

Solutions to these equations can be found as before for rays confined to a constant coordinate surface. Therefore putting $u = \text{constant}$, that is $\mathrm{d}u/\mathrm{d}s = 0$ in equations 5.24, and postulating that the medium is uniform in the z direction, that is, with $g_{33} = \text{constant} = 1$, equation 5.24c becomes

$$\eta \frac{\mathrm{d}z}{\mathrm{d}s} = \text{constant} = A \tag{6.22}$$

This is the generalization of Snell's law for a medium of this type. The constant A is specific to a given single ray and thus has different values for different rays in a pencil of rays. The precise parametric dependence of A is vital to the interpretation of the eventual mathematical results. Equations 5.24a and 5.24b reduce to

$$\frac{\eta g_{11}^{\frac{1}{2}}}{\Delta} \left(\frac{\mathrm{d}v}{\mathrm{d}s}\right)^2 (-f_v h_{vv} + h_v f_{vv}) = \frac{1}{g_{11}^{\frac{1}{2}}} \frac{\partial n}{\partial u} \tag{6.23a}$$

and

$$\frac{\mathrm{d}}{\mathrm{d}s} \left(\eta g_{22}^{\frac{1}{2}} \frac{\mathrm{d}v}{\mathrm{d}s}\right) = \frac{1}{g_{22}^{\frac{1}{2}}} \frac{\partial \eta}{\partial v} \tag{6.23b}$$

In addition, we now have

$$\mathrm{d}s^2 = g_{22}\mathrm{d}v^2 + \mathrm{d}z^2 \qquad \frac{\mathrm{d}}{\mathrm{d}s} = \frac{\partial}{\partial v} \frac{\mathrm{d}v}{\mathrm{d}s} + \frac{\partial}{\partial z} \frac{\mathrm{d}z}{\mathrm{d}s}$$

The ray constant A is now u-dependent.

We shall require the following identity

$$\frac{g_{11}}{\Delta g_{22}} (-f_v h_{vv} + h_v f_{vv}) = \frac{-1}{2} \frac{\partial}{\partial u} \log g_{22} \tag{6.24}$$

Table. Orthogonal cylindrical coordinates

(1) Circular polar coordinates: rays on cylinders

$$x = u \cos v \qquad y = u \sin v \qquad g_{22} = u^2$$

(2) Circular polar coordinates: rays in meridional planes

$$x = v \cos u \qquad y = v \sin u \qquad g_{22} = 1$$

(3) Exponential coordinates: rays on cylinders

$$z = e^w \qquad x = e^u \cos v \qquad y = e^u \sin v \qquad g_{22} = e^{2u}$$

(4)† Exponential coordinates: rays in meridional planes

$$z = e^{iw^*} \qquad x = e^v \cos u \qquad y = e^v \sin u \qquad g_{22} = e^{2v}$$

(5) Elliptic coordinates

$$z = a \cosh w \qquad x = a \cosh u \cos v \qquad y = a \sinh u \sin v$$

$$g_{22} = \sinh^2 u + \sin^2 v$$

(6) Logarithmic coordinates

$$z = a \log w \qquad x = a \log (u^2 + v^2) \qquad y = 2a \tan^{-1} (v/u)$$

$$g_{22} = 4a^2$$

(7) Tangent coordinates

$$z = \tan w \qquad x = \frac{\sin u \cos u}{\cos^2 u + \sinh^2 v}$$

$$y = \frac{\sinh v \cosh v}{\cos^2 u + \sinh^2 v} \qquad g_{22} = \frac{1}{\cos^2 u + \sinh^2 v^2}$$

(8)† $z \tan (iw^*) = -i \tanh w^*$ as for (7) with $u \leftrightarrow v$

(9) Inverse coordinates

$$z = 1/w \qquad x = \frac{u}{u^2 + v^2} \qquad y = \frac{v}{u^2 + v^2} \qquad g_{22} = \frac{1}{u^2 + v^2}$$

† The transformation $w \to iw^*$ is equivalent to the transformation $u \leftrightarrow v$.

where

$$g_{11} = f_u^2 + h_u^2 \qquad g_{22} = f_v^2 + h_v^2 \qquad \Delta = f_u h_v - f_v h_u$$

This result is obtained by applying the orthogonality relation of equation 5.20 and its derivative

$$f_u f_{vv} + f_v f_{uv} + h_u h_{vv} + h_v h_{uv} = 0$$

in turn to the left-hand side of equation 6.24.

Consider the first equation 6.23b. Multiplication by the factor $\eta g_{22}^{\frac{1}{2}}(dv/ds)$ gives, since $\partial \eta / \partial z = 0$

$$\eta g_{22}^{\frac{1}{2}} \frac{dv}{ds} \frac{d}{ds} \left(\eta g_{22}^{\frac{1}{2}} \frac{dv}{ds} \right) = \frac{\partial \eta}{\partial v} \frac{dv}{ds}$$

and thus

$$\left(\eta g_{22}^{\frac{1}{2}} \frac{dv}{ds}\right)^2 = \eta^2 + \text{constant} \tag{6.25}$$

the constant being u-dependent. From the definition of ds and equation 6.22 this constant is $-[A(u)]^2$, and consequently $|A(u)| < \eta$.

We now multiply equation 6.23a by g_{22} and use the identity in equation 6.24 and the result in equation 6.25 to get

$$-[A(u)]^2 \frac{1}{g_{22}} \frac{\partial g_{22}}{\partial u} = -2\eta \frac{\partial \eta}{\partial u}$$

or

$$\eta^2 = \frac{1}{g_{22}} \left\{ \int [A(u)]^2 \frac{\partial g_{22}}{\partial u} du \right\} + B(v) \tag{6.26}$$

with $B(v)$ as an arbitrary function of the integration. This is the general form of the refractive-index law for which a ray on a surface $u = $ constant is constrained to remain there.

The ray path can be obtained by integration of the ray equations separately; that is, from equation 6.22

$$\frac{dz}{ds} = \frac{A(u)}{\eta}$$

and from equation 6.25

$$g_{22} \frac{dv}{ds} = \frac{\{\eta^2 - [A(u)]^2\}^{\frac{1}{2}}}{\eta} \tag{6.27a}$$

Combining these gives

$$z = \int \frac{A(u)(f_v^2 + h_v^2)^{\frac{1}{2}}}{\{\eta^2 - [A(u)]^2\}^{\frac{1}{2}}} dv \tag{6.27b}$$

The ray is parallel to the z-axis at points where $dv/ds = 0$ (i.e. where $A(u) = \eta$), and transverse where $dz/ds = 0$ (where $\eta \to \infty$).

If the ray makes an angle ψ with a z-directed generator (Figure 6.12), then $dz/ds = \cos \psi$ and $g_{22}^{\frac{1}{2}}(dv/ds) = \sin \psi$, giving the 'starting' conditions for a ray from a source in the medium.

Although the analysis has been derived for the situation in which a ray is confined to a surface $u = $ constant, the simple substitution $u \leftrightarrow v$ in the transformation specifying the system (and in all subsequent results) gives the ray trajectory in the orthogonal system. Hence two orthogonal cross-sections of a ray pencil can be dealt with by the analysis given. Skew rays, of course, require the complete solution of the general equations.

Equations 6.26 and 6.27 can be solved for specific conditions of ray behaviour. The most important is coherence, that is, the phase length along each individual ray of a given pencil of rays must be equal. When the surface concerned is a closed

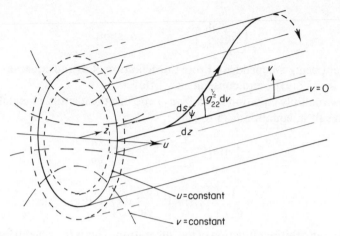

Figure 6.12 Rays in a general cylindrical coordinate system, confined to the surface u = constant

surface enclosing the axis, the ray paths will, in general, be spirals, encircling the axis, or they will be confined to a zone parallel to the axis. Coherence is maintained when rays of a particular pencil spiral with the same pitch.

For non-spiralling ray pencils, coherence is retained by continuous refocusing along the axis. Both these conditions are well known in the case of the circular polar-coordinate system: here it is extended to refractive-index variations in both the radial and the angular coordinates.

Any function $A(u)$ will give rays which spiral on the cylinder, in a system for which g_{22} is a function of u only. This can be seen by immediate substitution into equation 6.26. Subsequently dv/ds can only become zero in the limit of a completely uniform medium.

For example, in the circular system (1) of the foregoing Table with $g_{22} = u^2$, the function $A^2 = a - bu^2$ gives the commonly used 'parabolic' gradient of refractive index $\eta^2 = a - bu^2/2$ $(B(v) = 0)$. The helix angle of the spiral is then $\tan^{-1}\{2[(a/bu^2) - 1]\}$ which is u-dependent. A pencil of such rays will thus form a diffuse system, and coherence will not in general be retained. For weakly guiding media (small b) and small-diameter cylinders, the spiral pitch becomes large and diffusion of the rays can be kept small.

If, however, we dispense with the condition of regularity along the axis on the assumption that the guiding medium is a hollow tube, we can, in such a case, establish exact conditions for spiral rays from a flat pencil to retain coherence by making the pitch independent of the parameter u (but not, obviously, independent of ψ). This implies that the rays from an extended radial-line source would spiral as a ribbon with constant pitch, as shown in Figure 6.13.

Retaining $B(v) = 0$ in the first instance, then, for constant helix angle $p = \tan \psi$, the ratio of the two expressions in equation 6.23 becomes

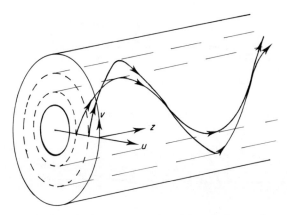

Figure 6.13 Spiralling band of rays on a tube with radially varying refractive index

$$\left\{\frac{\eta^2 - [A(u)]^2}{A(u)}\right\}^{\frac{1}{2}} = p \qquad (6.28)$$

and thus

$$\eta^2 = (p^2 + 1)[A(u)]^2 \qquad p > 0$$

Substitution into equation 6.26 gives for the circularly cylindrical system with $g_{22} = u^2$

$$(p^2 + 1)[A(u)]^2 = u^{-2} \int 2[A(u)]^2 u\, du$$

or

$$A(u) = au^{-p^2/(p^2 + 1)}$$

Such a solution is possible only where g_{22} is a function of u alone. Thus

$$\eta^2 = a^2(1 + p^2)u^{-2p^2/(p^2 + 1)}$$

The limit of a uniform medium and straight rays is attained as p tends to zero.

General functions of $B(v)$ can now be included. This has the effect of preventing the complete spiral from forming as the numerator in equation 6.23 has a zero value at some value of v. The ribbon of rays will fluctuate in a zone confined between these v values. Figure 6.14 shows the effect for arbitrarily chosen $B(v) = -2 \sin^2 v$.

With such an included function $B(v)$ the refractive-index law becomes

$$\eta^2 = a^2(1 + p^2)u^{-2p^2/(p^2 + 1)} + B(v)u^{-2} \qquad (6.29)$$

In this manner, isolated channels of rays can be formed on a single circular tubular fibre. In particular, for $p = 1$

$$\eta^2 = 2a^2u^{-1} + B(v)u^{-2}$$

It is now apparent that the same procedure can be adopted for any coordinate

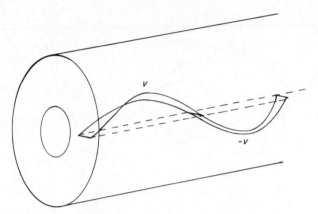

Figure 6.14 Frustrated spiral forms a coherent band within angular limits $\pm v$

system for which g_{22} is a function of u alone, and where the surfaces $u = $ constant are closed continuous cylinders containing the axis.

This generalizes the result of equation 6.29 to

$$\eta^2 = a^2(1 + p^2)[g_{22}(u)]^{-p^2/(1 + p^2)} + B(v)(g_{22})^{-1} \qquad (6.30)$$

Equation 6.27 is generally integrable by the method of Appendix III for those coordinate systems for which $g_{22} = f_v^2 + h_v^2$ is a constant or is a function of v only.

For rays not performing continuous spirals, the trajectory will in general fluctuate about a line generator of the cylinder. If we consider a pencil of rays from a source on the generator $v = 0$, then the angle made by the ray there is given by

$$\eta(u, v = 0) = A(u) \sec \psi$$

Also from equations 6.27, the ray will be parallel to the generators again at a value $v = v_1$, where

$$\eta(u, v = v_1) = A(u)$$

Since these occur on the surface $u = $ constant $= c$, we can designate them

$$\eta_{c,0} = A(c) \sec \psi$$

and

$$\eta_{c,1} = A(c)$$

On the surface itself, $\eta = \eta(c, v)$.

For the continuous coherent refocusing of a flat pencil of rays (flat in the sense of all rays being confined to the same surface), we require equation 6.27b to be independent of ψ on the surface $u = $ constant, i.e.

$$z = \int_0^{v_1} \frac{A(c)\{g_{22}^{\frac{1}{2}}(u = c)\}dv}{[\eta^2(c, v) - A^2(c)]^{\frac{1}{2}}} = \kappa \qquad (6.31)$$

where

$$\eta^2(c, v) = \frac{1}{g_{22}} \left[\int A^2(u) \frac{\partial g_{22}}{\partial u} \, du + B(v) \right]$$

evaluated at $u = c$.

For simplicity, we can write $\eta^2(c, v) = D^2(v)/g_{22}(u = c)$ and make the appropriate substitutions into equation 6.27 to give

$$\int_0^{v_1} \frac{\eta(c, 0) \cos \psi [g_{22}^{\frac{1}{2}}(u = c)] dv}{[D^2(v)/g_{22}(u = c) - \eta^2(c, 0) \cos^2 \psi]^{\frac{1}{2}}} = \kappa$$

or

$$\int_0^{v_1} \frac{D(0) \cos \psi [g_{22}(u = c)]}{[D^2(v) - D^2(0) \cos^2 \psi]^{\frac{1}{2}}} \, dv = \kappa \tag{6.32}$$

and where v_1 is the value of v for which

$$\cos \psi = \frac{D(v)}{D(0)}$$

that is, the ray is parallel to the axis.

Equation 6.32 can be integrated by the formal procedure of Appendix III, giving

$$D(v) = \text{sech} \left\{ \frac{\pi}{2\kappa} \int [g_{22}(u = c)] dv \right\}$$

and consequently

$$\eta(c, v) = [g_{22}^{-\frac{1}{2}}(u = c)] \, \text{sech} \left\{ \frac{\pi}{2\kappa} \int [g_{22}(u = c)] dv \right\}$$

and thus generally

$$\eta^2(u, v) = \frac{1}{g_{22}} \left(\int [A(u)]^2 \frac{\partial g_{22}}{\partial u} \, du + \text{sech}^2 \left\{ \frac{\pi}{2\kappa} \int [g_{22}(u = c)] dv \right\} \right) \tag{6.33}$$

The same analysis can be carried through for an orthogonal surface ($v = $ constant) by first transforming the surfaces by the substitution $u \leftrightarrow v$, deriving the results in equation 6.32 and then transforming back. Thus, two orthogonal-plane pencils of refocusing rays can exist in the cylindrical medium, not necessarily confined to the axis, since the surface $u = c$ and the source can be arbitrarily positioned. In this case $A(u)$ can be derived to symmetrize equation 6.33, with the result

$$\eta^2(u, v) = \frac{1}{g_{22}} \left\{ \text{sech}^2 \left[\frac{\pi}{2\kappa} \int g_{22}^{\frac{1}{2}}(u = c_1) dv \right] \right\}$$

$$+ \frac{1}{g_{11}} \left\{ \text{sech}^2 \left[\frac{\pi}{2\kappa} \int g_{11}^{\frac{1}{2}}(v = c_2) du \right] \right\} \tag{6.34}$$

for a source located on the intersection of the two surfaces $u = c_1$, $v = c_2$.

Coordinate systems can be chosen, which for small values of one coordinate can approximate to a rectangular strip, or even to a thin film guide, for which the above law, in its small-value approximation, would give the refractive-index profile for continuous self-focusing rays.

In certain cases, where the field can be expressed by a single component E vector, such as the perpendicular component of the field over a conducting ground plane, this refractive-index law can be converted to an equivalent height profile for a dielectric guiding strip.

7 Rays in Spherical and Axisymmetric Media

Ray-tracing in a non-uniform spherical medium provides a design method for many practical optical devices, mainly in the field of microwaves where transparent non-uniform refractive media can be constructed from foamed or loaded plastic materials. Early studies derived mainly from the work of biologists[30] who found that the focusing properties of the eyes of marine animals required such ray-tracing methods. New technologies applying to glass manufacture make it possible to design optical components with varying refractive index, and if a diametral cross-section of the spherical medium alone is considered then integrated optical components with specified focusing properties can be designed by ray-tracing methods.

However, besides being another example of the method for solving equations 5.18 and 5.30 some other theoretical consequences have been discovered similar to the inversion method given in previous chapters. There is a well-known analogy between optical rays and the dynamics of particles, both being describable by a similar form of least-action principle. The theoretical discoveries applying to optical rays therefore have an immediate connection with gauge transformation theory of the phase space of the dynamical system. In this chapter we shall deal with the practical design methods and transformations that are allowed and we shall consider the consequences of the transformation theories in the final chapter.

7.1 THE SPHERICALLY SYMMETRIC MEDIUM

The procedure is similar to that of the previous chapter. The ray equations become integrable when the ray is considered to be confined to a surface with a constant value of one coordinate, and for the first we take the spherically symmetrical medium with rays confined to the equatorial plane. The same analysis would be applicable to the *transverse* cross-section of the cylindrical systems of the previous chapter.

We therefore put $d\theta/ds = 0$, $\theta = \text{constant} = \pi/2$, and $\partial\eta/\partial\phi = 0$, $\partial\eta/\partial\theta = 0$ in equations 5.18, reducing them to

$$\frac{d}{ds}\left(\eta\frac{dr}{ds}\right) - \eta r\left(\frac{d\phi}{ds}\right)^2 = \frac{\partial\eta}{\partial r} \qquad (7.1a)$$

and

$$\frac{d}{ds}\left(\eta r \frac{d\phi}{ds}\right) + \eta \frac{dr}{ds}\frac{d\phi}{ds} = \frac{1}{r}\frac{\partial \eta}{\partial \phi} = 0 \tag{7.1b}$$

The ray is now given by $r = r(\phi)$, and

$$ds^2 = dr^2 + r^2 d\phi^2 \tag{7.2}$$

The consequence of equation 5.11

$$\text{curl } (\eta \hat{s}) = 0 \tag{7.3}$$

is that, in the complete refractive space, the rays are all closed loops. If we include the line at infinity, then closed loops in the projective plane includes curves such as parabolas and hyperbolas. This too exemplifies the dynamical analogue since equation 7.3 describes a conservative field, and hence one that derives from a potential function.

With circular symmetry equation 7.1b reduces to

$$\frac{d}{ds}\left(\eta r^2 \frac{d\phi}{ds}\right) = 0$$

or

$$\eta r^2 \frac{d\phi}{ds} = \text{constant} = A \text{ along a given ray} \tag{7.4}$$

Writing this as $\eta r r(d\phi/ds) = A$, we have

$$\eta r \sin \psi = A \tag{7.5}$$

where ψ is the angle between the radius vector to a point on the ray whose ray constant is A, and the tangent to the ray at that point (Figure 7.1). This generalizes Snell's law for a spherical medium and equation 7.5 is known as Bouguer's theorem.

Substitution of this result, and applying equation 7.2 to equation 7.1a, gives

$$d\phi = \frac{A dr}{r[r^2 \eta^2(r) - A^2]^{\frac{1}{2}}}$$

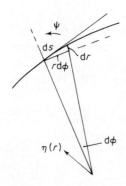

Figure 7.1 Ray in the meridional plane of a spherically symmetrical variable medium

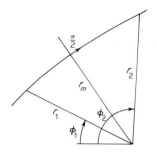

Figure 7.2 Turning value of the ray in the spherical medium occurs at the point where the ray tangent is perpendicular to the radius vector

and consequently

$$\phi_2 - \phi_1 = \int_{r_1}^{r_2} \frac{A\,dr}{r[r^2\eta^2 - A^2]^{\frac{1}{2}}} \tag{7.6}$$

This is the polar equation of the ray trajectory as shown in Figure 7.2. Due to the fact that r may be variously an increasing or decreasing function of ϕ, the range of integration has to be split at a value r_m where this changes. That occurs whenever the ray is at either its closest to the origin or the furthest away, and at both these points the ray will be parallel to the stratification. Hence ψ in equation 7.5 will be $\pi/2$ at these turning values, $r\eta(r)$ will equal A and the denominator in equation 7.6 will be zero. Thus the limit r_2 in the integral will be that value which makes the denominator zero.

If the boundary conditions are chosen such that $\phi_1 = 0$ when $r_1 = 1$, the polar equation of the ray is

$$\phi = \int_1^r \frac{A\,dr}{r[r^2\eta^2(r) - A^2]^{\frac{1}{2}}} \tag{7.7}$$

The solution of this equation for ϕ given as a function of r for the refractive index law $\eta(r)$ is again the method using Abel's integral given in Appendix III.

7.2 THE SPHERICAL SYMMETRIC LENSES

The archetypal non-uniform spherical lens is considered to be Maxwell's 'fish-eye'. In it, all the rays from a source on the perimeter of the unit sphere ($\phi = 0$ when $r = 1$) and at which point the refractive index is unity, are imaged into the diametrically opposite point. The result is the 'ideal lens' which images perfectly every point in the refractive space. Luneburg[25] proves that, apart from the trivial case of the plane mirror, it is the only optical device with this ability. There are several different ways by which this result can be obtained. Luneburg gives it as the result of a stereographic projection of the geodesics of the sphere on to the (diametral) plane. Alternatively it can be derived as a consequence of equation 7.3.[31] Here, and in what follows, we use the general principle of deciding *a priori* the ray path as a function of r and ϕ and integrating equation 7.7 by the method of Appendix III.

96

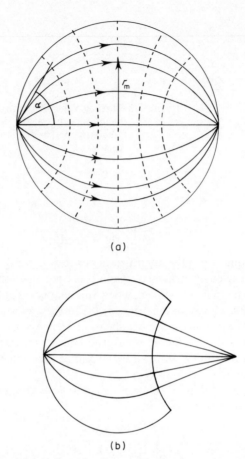

(a)

(b)

Figure 7.3 (a) Rays in Maxwell's fish-eye. The refractive index varies with the law
$\eta = 2/(1 + r^2)$ and refocuses all the rays at the diametrically opposite point
(b) The phase fronts, being circular, can define an exit surface with an external
focus

Since all rays come to a focus at the diametrically opposite point, the radius
vector has to turn through the angle $\phi = \pi$ for all values of A in equation 7.7.
Hence

$$\pi = 2 \int_1^{r_m} \frac{\sin \alpha \, dr}{r[r^2 \eta^2(r) - \sin^2 \alpha]^{\frac{1}{2}}} \tag{7.8}$$

where, as shown in Figure 7.3, r_m is the radius at which the ray becomes
horizontal. From the solution in Appendix III (equation III.7)

$$r\eta(r) = \text{sech} \{\log r\} \tag{7.9a}$$

and hence

$$\eta(r) = \frac{2}{1 + r^2} \tag{7.9b}$$

Substituting this result into equation 7.7 gives the ray trajectories as the coaxal system of circles, shown in Figure 7.3(a), with equations

$$r = (1 + \sin^2 \phi \cot^2 \alpha)^{\frac{1}{2}} + \sin \phi \cot \alpha \qquad (7.10)$$

Luneburg shows that this ideal focusing effect continues into the space external to the unit sphere and hence to regions where the refractive index is less than unity. Since the rays are a coaxal system of circles, the orthogonal system is also circular, and hence the phase fronts form circles with centres on the diameter containing the source. Thus if the lens is terminated by a surface coinciding with a spherical phase front, the ray will continue normal to that surface and come to a perfect focus at the point which is the centre of the spherical face (Figure 7.3(b)). In particular, if the terminating surface is itself a diametral plane the now hemispherical lens collimates all the rays into a beam focused at infinity (Figure 7.4).

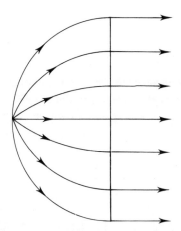

Figure 7.4 The hemispherical fish-eye lens focused at infinity

One generalization of the Maxwell fish-eye needs to be noted at this stage. If, instead of focusing all the rays to the diametrically opposite point as in Figure 7.3(a) the rays are all focused at a point $\phi = \pi/a$, then the left-hand side of equation III.7 in Appendix III becomes $2\pi/a$ and hence the refractive-index law

$$r\eta(r) = \text{sech} \left(\frac{a}{2} \log r \right)$$

or

$$\eta(r) = \frac{2r^{a-1}}{1 + r^{2a}} \qquad (7.11)$$

Obviously the system is no longer spherically symmetrical, and hence the ray pattern applies to diametral planes only or a source will be focused into a ring of foci. If confined to a single plane section, then if a is a whole number fraction of π, repeated foci will appear around the periphery and one such focus will coincide

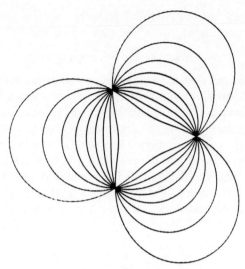

Figure 7.5 The generalized fish-eye refocusing at points on the unit circle—the ring resonator

with the source. The result would be a ring resonator. The ray pattern for this situation can be derived by the substitution of the law of equation 7.11 into the integral of equation 7.7, with the result shown in Figure 7.5

$$r^{-a} = (1 + \sin^2 a\phi \cot^2 \alpha) + \sin a\phi \cot \alpha \qquad (7.12)$$

The second class of non-uniform spherical lenses is again based upon the interior of the unit circle with, in the complete refractive space, perfect focusing of the rays at a point diametrically opposite to the source. In this second case, however, the rays are all parallel to the axis when they first meet the unit circle, at which point the refractive index is unity. Hence in real space, the rays all continue to be parallel to this direction after leaving the lens. The refractive law for such a lens was first derived by Luneburg,[25] and we can label this class of lens as the generalized Luneburg lens. For the general pattern of rays shown in Figure 7.6, all the rays from the source are required to arrive again at the unit circle where their directions will all be parallel to the angle $p\pi/2$ as shown. The similarity of this effect to the manner in which the fish-eye was generalized can be noted. Again ray spherical symmetry is lost, the result of a rotation being to create a cone of rays with half-angle $p\pi/2$. Hence, in the right-hand side of equation III.4, I is required to $\pi/2 + p\pi/4 - \alpha/2$, and thus the refractive-index law is given by

$$\{1 + [1 - r^2\eta^2(r)]^{\frac{1}{2}}\}^{p+1} = [r\eta(r)]^{p+2}r^{-2} \qquad (7.13)$$

If p is put equal to zero the result is Luneburg's collimating lens having both total symmetry and the refractive-index law

$$\eta^2(r) = 2 - r^2 \qquad (7.14)$$

The ray paths, from equation 7.7 using the substitution $r^2 = 1/v$, are ellipses

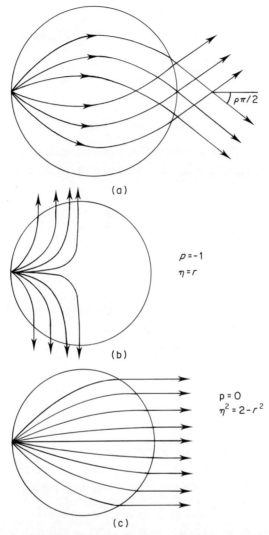

Figure 7.6 (a) Rays from the source on the unit circle separated into two beams at angles $\pm p/2$. The refractive law is obtained from

$$[1 + (1 - \eta^2 r^2)^{\frac{1}{2}}]^{p+1} = (\eta r)^{p+2} r^{-2}$$

(b) With $p = -1$ all rays are perpendicular to the axis and $\eta = r$
(c) With $p = 0$ all rays are parallel to the axis—the Luneburg lens with refractive law $\eta^2 = 2 - r^2$

(Figure 7.7) passing through the two diametrically opposite points with equations

$$r^2 = \sin^2 \alpha / [1 - \cos \alpha \cos (2\phi + \alpha)] \qquad (7.15)$$

The major axes are inclined at $\alpha/2$ to the diameter containing the source and of length $2\sqrt{2} \cos (\alpha/2)$, the minor axes are of length $2\sqrt{2} \sin (\alpha/2)$ and the foci are at

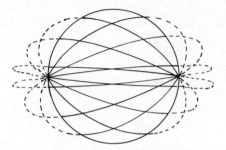

Figure 7.7 Complete elliptical rays in the Luneburg lens medium

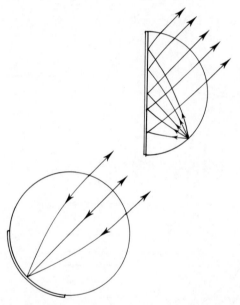

Figure 7.8 The bisected Luneburg reflector and the Luneburg lens retro-reflector

points $r = (\cos \alpha)^{\frac{1}{2}}$. The medium only extends to the maximum value $r^2 = 2$ from equation 7.14 and the refractive index is only larger than unity within the unit circle.

Adaptations of Luneburg's lens include the reflecting hemispherical lens, the retro-reflecting lens (Figure 7.8) and the lens with the source in the interior[32] (Figure 7.9). In the last of these the refractive-index law for a source at a distance f from the centre is given by Gutmann to be

$$\eta^2(r) = (1 + f^2 - r^2)/f^2 \qquad (7.16)$$

The rays in this lens are similar ellipses to those in Luneburg's lens but with increasing eccentricity for $f < 1$ as the source approaches the centre.

The Luneburg lens itself is also the ideal transformer between source radiation

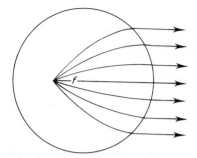

Figure 7.9 The Luneburg lens with internal source—the Gutman lens

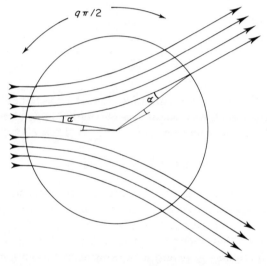

Figure 7.10 The radially variable beam divider. Power division can be achieved by displacing the input beam transversely from the axis

distributions of the form $\cos^m \theta$ and the standard lambda function radiation patterns. This is presented in Appendix IV.

One other ray pattern that can be directly dealt with is the situation shown in Figure 7.10, that is with a source at infinity, or a collimated beam, incident upon the lens which thereupon separates into a similar conical beam to the foregoing example. For all rays to make an angle $q\pi/2$ as shown, the substitution for the right-hand side of equation III.4 is $q\pi/2 - 2\alpha$, giving the refractive-index law

$$[1 + (1 - r^2\eta^2(r))]^{q-2} = [r\eta(r)]^2 r^{-2} \qquad (7.17)$$

When $q = 2$ the rays continue as straight lines through the lens and $\eta = 1$ as a consequence. When $q = 0$ the rays enter the lens, perform one orbit about the origin and are transmitted parallel to the incident direction. Equation 7.17 gives

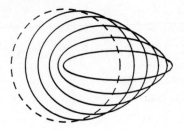

Figure 7.11 Elliptical rays in the Eaton lens,
$$\eta^2 = 2/r - 1$$

for this result the law

$$\eta^2 = \frac{2}{r} - 1 \tag{7.18}$$

which also results for $q = 4$.

Substitution of this law into equation 7.7 shows the rays to be ellipses all with a focus at the origin and all with major axis of length 2 lying along the diameter parallel to the direction of incidence (Figure 7.11). Their equations are

$$r = \sin^2 \alpha/(1 - \cos \alpha \cos \phi) \tag{7.19}$$

This lens was first derived by Eaton[33] as a perfect retro-reflector. Unfortunately the innermost ellipses require such high eccentricity that a principle of minimum curvature is violated and the resultant design is highly inefficient.

It should be noted that attempts to make a lens with refractive index higher than unity at the periphery have only led to approximate solutions, valid usually over a paraxial pencil at the source. The reason is that a refraction effect now occurs at the surface, making the solution by Abel's method of Appendix III inapplicable. One such is the lens of Toraldo di Francia[34] which has the law

$$r^2 \eta^2(r) = \eta^{1/m}(2 - \eta^{1/m}) \tag{7.20}$$

To maintain the completeness of the integration in Appendix III, the law of refraction at the surface has to be modified, in this case, to be $\kappa i = r$ instead of $\kappa \sin i = \sin r$, where κ is the refractive index at the surface and i and r are the incident and refracted angles.

Alternatively, rays which are hyperbolic could be specified[35] giving a refractive-index law

$$\eta^2(r) = 2\kappa + (\kappa^2 - 2^2\kappa)r^2 \tag{7.21}$$

valid over only a small range of angles at the source. Both these laws revert to Luneburg's law in the limit $\kappa = 1$.

7.3 TRANSFORMATIONS OF THE SPHERICAL LENSES

One obvious connection between the designs of the spherical non-uniform lenses is their general reliance on the use of the Abel integral method of Appendix III. It is therefore of interest to see what transformations exist that connect the lens designs or that can extend them.

The first such transformation is that relating to the phase fronts given by equation 5.6. Suppose as in equation 5.6 the surfaces $S(x, y, z) = $ constant are wavefronts in a refractive space $\eta(x, y, z)$, and $T(\lambda, \mu, v) = $ constant are wavefronts in a medium with refractive index $n(\lambda, \mu, v)$. These surfaces are connected by a Legendre transformation if

$$\frac{\partial S}{\partial x} = \lambda \qquad \frac{\partial S}{\partial y} = \mu \qquad \frac{\partial S}{\partial z} = v$$

$$\frac{\partial T}{\partial \lambda} = x \qquad \frac{\partial T}{\partial \mu} = y \qquad \frac{\partial T}{\partial v} = z$$

$$(7.22)$$

These can be collectively written $S + T = x\lambda + y\mu + zv$.

Since S and T are wavefronts, we have from equation 5.8

$$\left(\frac{\partial S}{\partial x}\right)^2 + \left(\frac{\partial S}{\partial y}\right)^2 + \left(\frac{\partial S}{\partial z}\right)^2 = \eta^2$$

$$\left(\frac{\partial T}{\partial \lambda}\right)^2 + \left(\frac{\partial T}{\partial \mu}\right)^2 + \left(\frac{\partial T}{\partial v}\right)^2 = n^2$$

$$(7.23)$$

We apply this to the fish-eye for which

$$\eta = \frac{2}{1 + r^2} = \frac{2}{1 + x^2 + y^2 + 2^2} = \left[\left(\frac{\partial S}{\partial x}\right)^2 + \left(\frac{\partial S}{\partial y}\right)^2 + \left(\frac{\partial S}{\partial z}\right)^2\right]^{\frac{1}{2}}$$

Then on transformation

$$(\lambda^2 + \mu^2 + v^2)^{\frac{1}{2}} = \frac{2}{1 + \left[\left(\frac{\partial T}{\partial \lambda}\right)^2 + \left(\frac{\partial T}{\partial \mu}\right)^2 + \left(\frac{\partial T}{\partial v}\right)^2\right]^{\frac{1}{2}}}$$

or

$$n^2 = \frac{2}{\rho} - 1$$

which is the refractive-index law, equation 7.18, of the lens of Eaton.

It is also to be noted that this same result can be effected by the transformation $\eta \rightarrow r$ and $r \rightarrow \eta$, leaving $r\eta(r)$ invariant. It does not always result in an explicit form for the refractive index, and once effected the rays have to be recalculated with the new refractive law.

The second transformation is of a more fundamental nature. It could be anticipated, from the fact that the only completely conformal mapping of a three-dimensional space onto itself (apart from the elementary translation, rotation or magnification) is an inversion,[36] that this would be the transformation most relevant to spherically symmetrical problems.

As inversion leaves angles invariant, Bouguer's theorem for a spherical medium (equation 7.5)

$$r\eta(r) \sin \alpha = r_0\eta(r_0) \sin \alpha_0$$

(the suffix zero indicating the source condition for a ray with tangent making an angle α with the radius vector from the origin), would imply that the proper function for the transformation process is not the refractive index itself, but the function

$$r\eta(r) = f(r)$$

and this accords with the invariance of this function under the Legendre transformation.

We therefore define a generalized inversion[37]

$$(r, \phi) \rightarrow (R, \Phi)$$

such that the refractive index transforms by

$$r\eta(r) = f(r) \rightarrow r\bar{\eta}(r) = f(1/r^n) \qquad (7.24)$$

That is, in general, $r \rightarrow 1/R^n$ in the coordinate space, whereas the refractive index alone transforms by

$$\eta(r) = f(r)/r \rightarrow \bar{\eta}(R) = f(1/R^n)/R$$

For general values of n, the optical line element is scaled by a factor n. For a ray confined to the equatorial plane of the medium

$$\eta^2(r)ds^2 = f^2(r)\{dr^2/r^2 + d\phi^2\}$$

which transforms to

$$f^2(1/R^n)(n^2 dR^2/R^2 + d\phi^2)$$

Thus, if $\phi \rightarrow n\Phi$, we obtain

$$\eta^2(r)ds^2 = n^2 f^2(1/R^n)\{dR^2 + R^2 d\Phi^2\}/R^2$$

$$= n^2\bar{\eta}^2(R)\overline{ds}^2$$

We can now show that this formulation transforms the rays in one medium directly into rays in the transformed medium. If a ray at a fixed point (r_0, ϕ_0) makes an angle α_0 with the radius vector at that point, the ray constant κ is given by Bouguer's theorem to be

$$\kappa = r_0\eta(r_0) \sin \alpha_0 = f(r_0) \sin \alpha_0$$

the ray trajectory will then be given by equation 7.7

$$\phi - \phi_0 = \int_r^{r_0} \frac{-\kappa dr}{r[f^2(r) - \kappa^2]^{\frac{1}{2}}}$$

This will transform by the method given to

$$n\Phi - n\Phi_0 = \int_{1/R^n}^{1/R_0^n} \frac{n\bar{\kappa}\mathrm{d}R}{R\left[f^2\left(\frac{1}{R^n}\right) - \bar{\kappa}^2\right]^{\frac{1}{2}}}$$

and hence

$$\Phi - \Phi_0 = \int_{1/R^n}^{1/R_0^n} \frac{\bar{\kappa}\mathrm{d}R}{R[R^2\bar{\eta}^2(R) - \bar{\kappa}^2]^{\frac{1}{2}}} \tag{7.25}$$

where $\bar{\kappa} = f(1/R_0^n)\sin{(\pi - \alpha_0)}$ is the transform of κ.

Equation 7.25 is the ray equation for rays in the medium $\bar{\eta}(R)$ connecting points with coordinate $1/R_0^n$ and $1/R^n$.

Thus, points (r, ϕ) on rays in the first medium transform directly into points $(1/r^n, n\phi)$ in the second medium. The change of sign in equation 7.25 arises because a ray in one medium, in which r increases with ϕ, will transform into a ray in the new medium in which $1/r^n$ decreases with increasing ϕ, or vice versa.

The special case of the above for $n = 1$ extends the argument from the equatorial plane to any skew ray trajectory in a three-dimensional space, and leaves the optical line element invariant.

We then have

$$\eta^2(r)\mathrm{d}s^2 = f^2(r)(\mathrm{d}r^2/r^2 + \mathrm{d}\phi^2 + \sin^2\phi\mathrm{d}\psi^2)$$

With $r \to 1/R$ and $\phi \to \Phi$ this gives (ψ is invariant)

$$\eta^2(r)\mathrm{d}s^2 = f^2(1/R)(\mathrm{d}R^2/R^2 + \mathrm{d}\Phi^2 + \sin^2\Phi\mathrm{d}\psi^2)$$
$$= \bar{\eta}^2(R)\overline{\mathrm{d}s^2}$$

Applying this transformation to the lenses so far discussed gives an entire new class of refractive media in which the rays are the true geometrical inversions of the rays of the original lenses.

(i) The medium with $r\eta(r)$ constant

This law is obviously self-inverse. In the medium with $\eta(r) = 1/r$ the rays are equiangular spirals, with pole at the origin, which are likewise self-inverse with respect to an inversion in the pole (see Appendix II.1).

(ii) The Maxwell fish-eye

From the refractive law $\eta(r) = 2/(1 + r^2)$ we obtain $r\bar{\eta}(r) = (2/r)/(1 + 1/r^2)$ or $\bar{\eta}(r) = 2/(1 + r^2)$. Hence this law is also invariant. In this case the inversion transforms a circle of the coaxal system in Figure 7.3(a) into its reflection in the axis—that is, a ray with source angle α into the ray with source angle $\pi - \alpha$.

106

(iii) *The Luneburg lens*

The refractive law is given by

$$\eta(r) = (2 - r^2)^{\frac{1}{2}}$$

which on inversion becomes

$$\bar{\eta}(r) = (2r^2 - 1)^{\frac{1}{2}}/r^3$$

and hence the ellipses given by equation 7.15 become the Cassinian ovals (Figure 7.12)

$$r^2 = [1 - \cos \alpha \cos (2\phi + \alpha)]/\sin^2 \alpha$$

Figure 7.12 The inversion of the elliptical rays of the Luneburg lens into Cassinian ovals

(iv) *The Eaton lens*

The refractive law for this is

$$\eta(r) = \left(\frac{2}{r} - 1\right)^{\frac{1}{2}}$$

and hence on inversion we obtain

$$\bar{\eta}(r) = (2r - 1)^{\frac{1}{2}}/r^2$$

The elliptical rays of equation 7.19 thus invert to become limaçons of Pascal with equations

$$r = (1 - \cos \alpha \cos \phi)/\sin^2 \alpha$$

as shown in Figure 7.13.

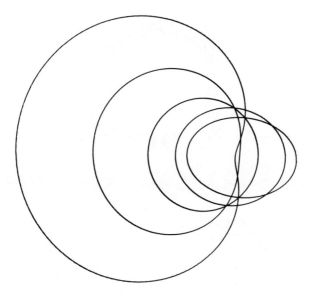

Figure 7.13 Inversion of the elliptical rays of the Eaton lens into limaçons of Pascal

(v) The uniform medium

This has rays which are straight lines. The law $\eta(r) = $ constant transforms to $\bar{\eta}(r) = 1/r^2$, which therefore has rays which are circles passing through the centre (for a source not at the origin). This can be simply confirmed.

It is elementary to show that, for this case $n = 1$, the repeat of an inversion reproduces the original medium.

The effect of an inversion with $n \neq 1$ is complicated by the factor multiplying ϕ. Thus, the ray in one medium can be 'compressed' in angle for $n > 1$ or extended, even 'wrapped around' the origin in a new medium with $n < 1$.

For example, with $n = 2$ the transformation is $r \to 1/r^2$, applied to the refractive index

$$r\eta(r) = r(2 - r^2)^{\frac{1}{2}}$$

of the Luneburg lens results in

$$r\bar{\eta}(r) = (2 - 1/r^4)^{\frac{1}{2}}/r^2$$

For a ray from the same source making the angle $(\pi - \alpha)$ with the axis, the trajectory will therefore be

$$\phi = \int_1^r \frac{-\sin \alpha \, dr}{r \left[\frac{1}{r^4} \left(2 - \frac{1}{r^4} \right) - \sin^2 \alpha \right]^{\frac{1}{2}}}$$

which integrates now, with the substitution $r^2 = v$, to give the identical integrand

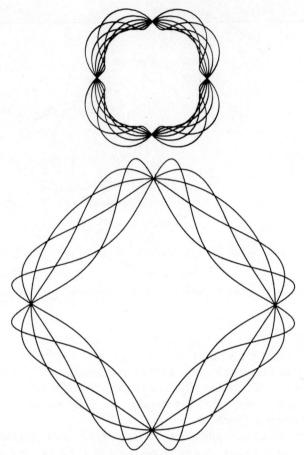

Figure 7.14 (a) Inversion of the rays of the Luneburg lens for $n = 2$
(b) Inversion of rays in the Luneburg lens for $t = 2$

as arises from equation 7.14, with the result

$$r^4 = [1 - \cos \alpha \cos (4\phi + \alpha)]/\sin^2 \alpha$$

This, as shown in Figure 7.14, is a four-fold version of the Cassinian oval of Figure 7.12. Similarly, $n = 3$ would result in a six-fold version.

The analysis for fractional values of n follows identical lines. The transformed refractive indices and ray paths for the modified Luneburg lens are

$$n = \tfrac{1}{2}: \quad \bar{\eta}(r) = (2 - 1/r)^{\frac{1}{2}}/r^{\frac{3}{2}}$$

$$r = [1 - \cos \alpha \cos (\phi + \alpha)]/\sin^2 \alpha$$

$$n = \tfrac{1}{3}: \quad \bar{\eta}(r) = (2 - 1/r^{\frac{2}{3}})^{\frac{1}{2}}/r^{\frac{4}{3}}$$

$$= \{[1 - \cos \alpha \cos (\tfrac{2}{3}\phi + \alpha)]/\sin^2 \alpha\}^{\frac{3}{2}}$$

$$0 < \phi < 4\pi$$

It can be shown that the reciprocal of an inversion of degree n is an inversion of degree $1/n$.

Composite media can be considered in which the interior of the unit sphere consists of a given known medium and the exterior consists of an inverse, both media having *a priori* defined ray paths. Regions hitherto inaccessible to rays, for instance $r > \sqrt{2}$ in the Luneburg lens medium, will in this way become usable as lens media.

The transformation provides a direct proof of the refocusing properties of the rays in the generalized fish-eye medium given by equation 7.11. We have for the Maxwell fish-eye

$$r\eta(r) = \frac{2r}{1 + r^2}$$

and hence for an inversion of degree n

$$r\bar{\eta}(r) = \frac{2/r^n}{1 + 1/r^{2n}}$$

or

$$\bar{\eta}(r) = \frac{2r^{n-1}}{r^{2n} + 1} \quad \text{the generalized medium}$$

The rays in Maxwell's fish-eye are the system of coaxal circles with parameter α given by (equation 7.10)

$$r = \{1 + \sin^2 \phi \cot^2 \alpha\}^{\frac{1}{2}} + \sin \phi \cot \alpha$$

Hence the rays in the generalized medium will be, as in equation 7.12

$$r^{-n} = \{1 + \sin^2 n\phi \cot^2 \alpha\}^{\frac{1}{2}} + \sin n\phi \cot \alpha$$

and this gives the $2n$-fold refocusing at points around the unit circle shown in Figure 7.5.

Inversions for negative values of n are permissible, except that the condition governing the change in sign of equation 7.25 no longer applies. Hence, for negative n the rays transform from $F(r, \phi) = 0$ to $F(1/r^n, -n\phi) = 0$. The identity transformation is then given by $n = -1$.

Inversions of different degree may be compounded. In general, we then have for two consecutive inversions of degree n and m

$$r\eta(r) = f(r) \rightarrow r\bar{\eta}(r) = f(1/r^n) \rightarrow r\bar{\bar{\eta}}(r) = f(r^{nm}) \tag{7.26}$$

and the rays with equation $F(r, \phi) = 0$ become rays

$$F(r^{nm}, nm\phi) = 0 \tag{7.27}$$

It can be seen that this is equivalent to the single inversion of degree $-nm$. This process produces a medium once again 'realistic' in the interior of the unit circle. Putting nm equal to the single transformation t, say, results in a further set of refractive scaling factors. For example, with $nm = t$, the transformation of equation 7.26 for the Luneburg lens results in

$$r\eta(r) = r(2 - r^2)^{\frac{1}{2}} \rightarrow r\bar{\eta}(r) = r^t(2 - r^{2t})^{\frac{1}{2}}$$

$$r^2\eta^2(r) = r^{2t}(2 - r^{2t}) \tag{7.28}$$

which is precisely that of the Toraldo lens of equation 7.20 under the Legendre transformation $\eta \rightarrow r, r \rightarrow \eta$. The rays can be derived directly from equation 7.15 to be (Figure 7.14(b))

$$r^{2t} = \sin^2 \alpha / [1 - \cos \alpha \cos (2t\phi + \alpha)] \tag{7.29}$$

7.4 RAYS IN AN AXISYMMETRIC COORDINATE SYSTEM

The ray equations for the axisymmetric coordinate system were given in Chapter 5, equations 5.29 and 5.30. A special case, as we shall see, is the spherical system again with the rays confined to the surface of the sphere. The problem is of interest more in the dynamical sense where the refractive index is analogous to a potential field and the rays to particle trajectories. The resulting trajectories have elements in common with the winds and tides prevailing, where the rotation of the Earth can be assumed to produce the potential field.

Equations 5.29 and 5.30 are too general for complete solution. Assuming the rays to be confined to surfaces $\alpha = $ constant allows us to put $\mathrm{d}\alpha/\mathrm{d}s = 0$ and obtain

$$-\eta \frac{K}{H_\beta} (F_\beta^2 + H_\beta^2)^{\frac{1}{2}} \frac{\mathrm{d}\beta}{\mathrm{d}s} \frac{\mathrm{d}}{\mathrm{d}s} \left[\frac{H_\alpha(F_\beta^2 + H_\beta^2)^{\frac{1}{2}}}{K} \right] - \eta F F_\alpha \left(\frac{\mathrm{d}\psi}{\mathrm{d}s} \right)^2 = \frac{\partial \eta}{\partial \alpha} \tag{7.30}$$

$$\frac{\mathrm{d}}{\mathrm{d}s} \left[\eta(F_\beta^2 + H_\beta^2)^{\frac{1}{2}} \frac{\mathrm{d}\beta}{\mathrm{d}s} \right] - \eta \frac{FF_\beta \left(\dfrac{\mathrm{d}\psi}{\mathrm{d}s} \right)^2}{(F_\beta^2 + H_\beta^2)^{\frac{1}{2}}} = \frac{1}{(F_\beta^2 + H_\beta^2)^{\frac{1}{2}}} \frac{\partial \eta}{\partial \beta} \tag{7.31}$$

$$\frac{\mathrm{d}}{\mathrm{d}s} \left(\eta F \frac{\mathrm{d}\psi}{\mathrm{d}s} \right) + \eta F_\beta \frac{\mathrm{d}\beta}{\mathrm{d}s} \frac{\mathrm{d}\psi}{\mathrm{d}s} = 0 \tag{7.32}$$

Since F, H and K and related derivatives as well as η are themselves ψ-independent

$$\frac{\mathrm{d}}{\mathrm{d}s} \equiv \frac{\partial}{\partial \beta} \frac{\mathrm{d}\beta}{\mathrm{d}s} \tag{7.33}$$

$$\mathrm{d}s^2 = (F_\beta^2 + H_\beta^2)\mathrm{d}\beta^2 + F^2\mathrm{d}\psi^2 \tag{7.34}$$

Substituting for $\mathrm{d}s^2$ in equation 7.30, we obtain

$$-\eta \left\{ \frac{K}{H_\beta} (F_\beta^2 + H_\beta^2)^{\frac{1}{2}} \frac{\partial}{\partial \beta} \left[\frac{H_\alpha(F_\beta^2 + H_\beta^2)^{\frac{1}{2}}}{K} \right] \mathrm{d}\beta^2 + FF_\alpha\mathrm{d}\psi^2 \right\}$$

$$= \frac{\partial \eta}{\partial \alpha} [(F_\beta^2 + H_\beta^2)\mathrm{d}\beta^2 + F^2\mathrm{d}\psi^2] \tag{7.35}$$

The coefficients of $d\beta^2$ and $d\psi^2$ on each side of this equation are equal, that is

$$-\eta F F_\alpha = F^2 \frac{\partial \eta}{\partial \alpha} \qquad (7.36)$$

and

$$-\eta \frac{K}{H_\beta} (F_\beta^2 + H_\beta^2)^{\frac{1}{2}} \frac{\partial}{\partial \beta} \left[\frac{H_\alpha (F_\beta^2 + H_\beta^2)^{\frac{1}{2}}}{K} \right] = \frac{\partial \eta}{\partial \alpha} (F_\beta^2 + H_\beta^2) \qquad (7.37)$$

Equation 7.36 is integrable and gives

$$\eta = \frac{A(\beta)}{F(\alpha, \beta)} \qquad (7.38)$$

where $A(\beta)$ is an arbitrary function of the integration. Dividing equation 7.37 by equation 7.36, and performing the differentiations, results, after some manipulation, in the solubility condition

$$F\{F_{\alpha\beta} H_\alpha - H_{\alpha\beta} F_\alpha\} + H_\beta \{F_\alpha^2 + H_\alpha^2\} = 0 \qquad (7.39)$$

It can be noted that, had the ray been assumed to be confined to a surface of constant β, the analysis above, and the solubility condition, would have been arrived at in a similar manner subject to the substitutions $\alpha \leftrightarrow \beta$.

The relation for η in equation 7.38 acts as an integrating factor for equation 7.32, with the result

$$\frac{d}{ds} \left[A(\beta) F(\alpha, \beta) \frac{d\psi}{ds} \right] = 0$$

or

$$A(\beta) F(\alpha, \beta) \frac{d\psi}{ds} = D \qquad (7.40)$$

a constant for the ray trajectory. This is a three-dimensional generalization of Snell's law and contains all the previously known generalizations, such as Bouguer's theorem.

Substitution of this result, and that for η, in equation 7.31 results in

$$\frac{d^2\beta}{ds^2} + \left(\frac{d\beta}{ds} \right)^2 \frac{(F_\beta F_{\beta\beta} + H_\beta H_{\beta\beta})}{(F_\beta^2 + H_\beta^2)} - \frac{D^2 A'(\beta)}{A^2(\beta)(F_\beta^2 + H_\beta^2)} = 0$$

or (by inspection !)

$$\frac{d\beta}{ds} = \left\{ \frac{1 - \dfrac{D^2}{[A(\beta)]^2}}{F_\beta^2 + H_\beta^2} \right\}^{\frac{1}{2}} \qquad (7.41)$$

The condition for solubility appears to be highly restrictive, in that nearly all the common forms of axisymmetric and toroidal coordinate systems available[38] do not satisfy the criterion of equation 7.39.

Quite general orthogonal axisymmetric coordinates can be obtained from the complex mapping

$$G(x + iz) = F(\alpha, \beta) + iH(\alpha, \beta)$$

and subsequent rotation about the z axis. These can be generalized still further into toroidal coordinate systems by translation along the positive x axis before rotation.

The simplest coordinate system so far found to satisfy equation 7.39 is that defined by

$$F(\alpha, \beta) = f(\alpha) \cos [g(\beta)]$$

$$H(\alpha, \beta) = f(\alpha) \sin [g(\beta)]$$

for general functions f and g. In this, and in most of the following, sine and cosine functions are found to be interchangeable, provided consistency is maintained. However, surfaces of constant α are spheres centred on the origin with radius $f(\alpha)$ and, consequently, the simplest permissible solutions are

$$F(\alpha, \beta) = f(\alpha) \cos \beta$$

$$H(\alpha, \beta) = f(\alpha) \sin \beta$$

With this definition the factor $F_\beta F_{\beta\beta} + H_\beta H_{\beta\beta}$ in equation 7.31 becomes zero. (This would also have been the case if, for $g(\beta)$, we had had $g''(\beta) = 0$ giving $g(\beta) = n\beta$, which is only a slight generalization and will be included subsequently.)

The final versions of equations 7.40 and 7.41 thus become

$$A(\beta)f(\alpha) \cos \beta \frac{d\psi}{ds} = D \qquad (7.42a)$$

$$\frac{d\beta}{ds} = \left[1 - \frac{D^2}{A(\beta)^2}\right]^{\frac{1}{2}} \qquad (7.42b)$$

with

$$\eta = \frac{A(\beta)}{f(\alpha) \cos \beta}$$

Equation 7.30 now becomes

$$-\eta f(\alpha)f'(\alpha)\left(\frac{d\beta}{ds}\right)^2 - \eta f(\alpha)f'(\alpha) \cos^2 \beta \left(\frac{d\psi}{ds}\right)^2 = \frac{\partial \eta}{\partial \alpha} \qquad (7.43)$$

with a solution $f'(\alpha) = 0$; $\eta'(\alpha) = 0$, so that the refractive index in the radial direction is uniform, as required for a ray to be confined to a spherical surface.

There remains at our disposal the arbitrary function $A(\beta)$. Equations 7.42 constitute the equations for the ray path in this medium and, thus the choice of the function $A(\beta)$ must be such that it contains no zeros in the range of β required, or for which $d\beta/ds$ is real. We will take β to be the angle of latitude on the sphere and, hence, its range will be $-\pi/2 \leqslant \beta \leqslant \pi/2$.

In the cases where $A(\beta)$ becomes infinite, the refractive index does too and we resort to the convention that rays terminate (or originate) there in a direction at right-angles to the local stratification of the refractive index. For real physical situations, regions of 'realistic' refractive index only will be considered, and thus ray paths near singularities such as infinities of the refractive index or at the poles of the sphere will be purely hypothetical.

Equation 7.42 can now be integrated to give the angle of travel about the axis, as follows

$$d\psi = \int \frac{D}{A(\beta) \cos \beta} \, ds$$

on a sphere $\alpha = $ constant, and hence

$$\psi_1 - \psi_0 = \int_{\beta_0}^{\beta_1} \frac{D d\beta}{\cos \beta [\{A(\beta)\}^2 - D^2]^{\frac{1}{2}}} \tag{7.44}$$

where β_1 and β_0 are the values of the polar latitude of the end-points of a ray.

We consider the pencil of rays from a point source on the equator confined to the surface of the sphere. This situation produces the boundary condition for the solution of the ray integral equation.

In order to contain the ray within a zone about the equator, it is intuitively obvious that the refractive index should have a maximum value there, and so we can anticipate a function $A(\beta)$ of the form

$$A(\beta) = AB(\beta) \tag{7.45}$$

where $B(\beta)$ has a maximum value of unity when $\beta = 0$.

At the source, therefore (Figure 7.15)

$$\left.\frac{d\psi}{ds}\right|_{\beta=0} = \frac{D}{A}$$

$$\left.\frac{d\beta}{ds}\right|_{\beta=0} = \left(1 - \frac{D^2}{A^2}\right)^{\frac{1}{2}}$$

That is, $D \leqslant A$ and the ray makes an angle with the equator at the source of

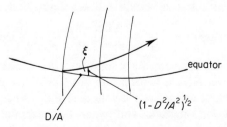

Figure 7.15 Rays on the surface of a sphere with angular variation of refractive index

$D/A = \cos \xi$. The ray becomes parallel to the equator on angles (ψ, β), where

$$\frac{d\beta}{ds} = 0 = 1 - \left[\frac{D^2}{A^2 B^2(\beta)}\right]$$

and turns to a line of longitude where $d\psi/ds = 0$.

From equation 7.42 this occurs at values for which $A(\beta)$ is infinite, that is the refractive index also becomes infinite.

From equation 7.45, for a ray commencing at the equator making an angle ξ there, at longitude $\psi_0 = 0$, the ray will be parallel to the equator at longitude

$$\psi_p = \int_0^{\beta_0} \frac{\dfrac{D}{A} \, d\beta}{\cos \beta \left[B(\beta)^2 - \dfrac{D^2}{A^2}\right]^{\frac{1}{2}}}$$

$$= \int_0^{\beta_0} \frac{\cos \xi \, d\beta}{\cos \beta [B(\beta)^2 - \cos^2 \xi]^{\frac{1}{2}}} \tag{7.46}$$

where β_0 is the solution of $d\beta/ds = 0$ or, from equation 7.44

$$B(\beta_0) = \cos \xi$$

The ray will reach the equator again at longitude $2\psi_p$. Thus, if ψ_p can be made independent of ξ, all the rays will come to a focus at the same point on the equator. In particular, if $2\psi_p$ is a whole-number fraction of 2π, one of these foci will coincide with the original source. For this to occur, therefore, we require a solution of the integral equation

$$\psi_p = \int^{\beta_0} \frac{\cos \xi \, d\beta}{\cos \beta [B^2(\beta) - \cos^2 \xi]^{\frac{1}{2}}} = \frac{\pi}{v} \tag{7.47}$$

with v preferably, but not essentially, an integer. Not surprisingly, the solution can be obtained by identical methods to the standard treatment of rays in a purely spherically symmetrical medium, given in Appendix III. As shown there this results in

$$B(\beta) = \frac{2 \cos^{v/2} \beta}{(1 + \sin \beta)^{v/2} + (1 - \sin \beta)^{v/2}} \tag{7.48}$$

This function for various values of v is shown in Figure 7.16.

The resulting distribution of refractive index obtained is

$$\cos \beta \eta(\beta) = \frac{2A \cos^{v/2} \beta}{(1 + \sin \beta)^{v/2} + (1 - \sin \cdot \beta)^{v/2}} \, \eta(\alpha)$$

With $v = 0$ we obtain the elementary solution $A(\beta) = $ constant. This implies a variation of refractive index $\eta = \sec \beta$ and $\cos \beta (d\psi/ds) = $ constant as well as $d\beta/ds = $ constant. These are respectively the sine and cosine of the angle ξ made by the ray at a line of longitude, and, when both remain constant, give the loxodrome path on the sphere (Figure 7.17). On a Mercator's projection these are

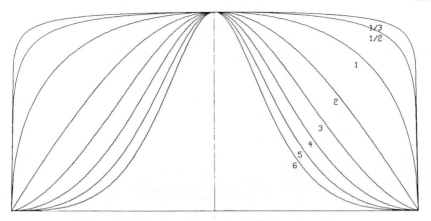

Figure 7.16 The law of equation 7.48 for values of v

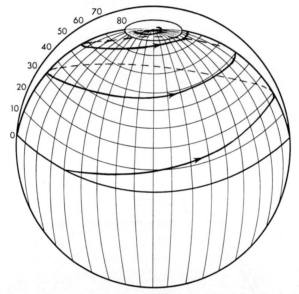

Figure 7.17 Loxodrome ($v = 0$) rays on the sphere with refractive index $\sec \beta$

rhumb lines, that is straight lines crossing the meridians at a fixed angle.

With $v = 1$ the rays are parallel to the equator at the longitude diametrically opposed to the source, and hence circle the pole once (Figure 7.18).

With $v = 2$, $B(\beta) = \cos \beta$ and the result confirms the standard result (see also equation III.3)

$$\int_0^\xi \frac{\cos \xi d\beta}{\cos \beta (\cos^2 \beta - \cos^2 \xi)^{\frac{1}{2}}} = \frac{\pi}{2}$$

The rays are then great circles coming to a focus at the point at the equator diametrically opposite the source (Figure 7.19). The refractive index in this case is uniform.

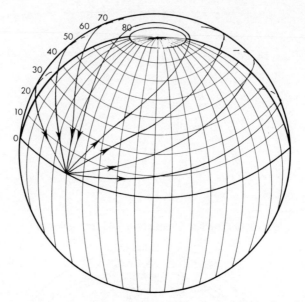

Figure 7.18 *v* = 1 rays are parallel to the equator at the longitude directly opposite the source

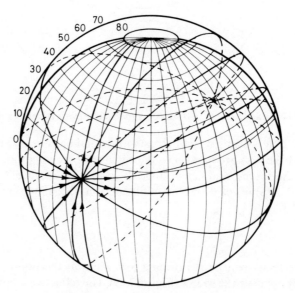

Figure 7.19 *v* = 2 rays are great circles

$v = 4$ gives the result

$$\int_0^{\sin^{-1}(\tan\xi/2)} \frac{\cos \xi d\beta}{\cos \beta \left(\dfrac{\cos^4 \beta}{1 + \sin^2 \beta} - \cos^2 \xi \right)^{\frac{1}{2}}} = \frac{\pi}{4}$$

which is capable of confirmation by the application of complete elliptic integrals of the first and third kinds, but higher values of v require hyperelliptic techniques or computational methods. Rays for other values of v are shown in Figure 7.20.

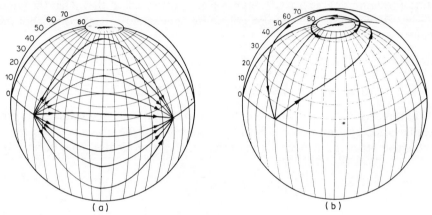

(a)   (b)

Figure 7.20 Surface rays for (a) $v = 3$ and (b) $v = \frac{1}{2}$

All other functions of $A(\beta)$ give diffuse ray systems: some, particularly those arising from $A(\beta) = \cos^n \beta$, are still confined to a zone about the equator.

The results of the previous section can be generalized for separable solutions. That is, letting

$$F(\alpha, \beta) = f_1(\alpha)f(\beta)$$

$$H(\alpha, \beta) = h_1(\alpha)h(\beta) \tag{7.49}$$

the orthogonality condition and equation 7.39 require

$$f_1(\alpha) \equiv h_1(\alpha) = \text{constant}$$

$$f^2(\beta) + h^2(\beta) = 1$$

After substitution for $d\psi/ds$, equation 7.31 then becomes

$$\frac{d^2\beta}{ds^2} + \left(\frac{d\beta}{ds} \right)^2 \left\{ \frac{f''(\beta)}{f'(\beta)} + \frac{f'(\beta)f^2(\beta)}{f(\beta)[1 - f^2(\beta)]} \right\} - \frac{A'(\beta)}{A^3(\beta)} D^2 \frac{(1 - f^2(\beta))}{f'^2(\beta)} = 0$$

$$\frac{d\beta}{ds} = \frac{[1 - f^2(\beta)]^{\frac{1}{2}}}{f'(\beta)} \left[1 - \frac{D^2}{A^2(\beta)} \right]^{\frac{1}{2}}$$

The resultant functions $B(\beta)$ for repeated focusing are then given by

$$\int_0^{\beta_0} \frac{f'(\beta)}{f(\beta)[1 - f^2(\beta)]^{\frac{1}{2}}} \frac{\frac{D}{A} d\beta}{[B^2(\beta) - D^2/A^2]^{\frac{1}{2}}} = \frac{\pi}{\nu}$$

or

$$B(\beta) = \frac{2[f(\beta)]^{\nu/2}}{\{1 - [1 - f^2(\beta)]^{\frac{1}{2}}\}^{\nu/2} + \{1 + [1 - f^2(\beta)]^{\frac{1}{2}}\}^{\nu/2}} \tag{7.50}$$

In fact these functions are not as general as first appears but are scale transformations of the β axis. For example, if $f(\beta) = \operatorname{sech} \beta$, then β has the range $(-\infty, \infty)$, where ∞ refers to one pole of the sphere and $-\infty$ the other, but the functional form of the refractive index is unaltered.

7.5 RAYS IN AN ANGULAR VARIABLE MEDIUM

Referring again to the equations for the spherical polar coordinate system (equations 5.18), for rays confined to a diametral plane $d\phi/ds = 0$ and for a refractive index independent of both the r and ϕ coordinates. They reduce to

$$\frac{d}{ds}\left(\eta \frac{dr}{ds}\right) - \eta r \left(\frac{d\theta}{ds}\right)^2 = 0 \tag{7.50}$$

$$\frac{d}{ds}\left(\eta r \frac{d\theta}{ds}\right) + \frac{dr}{ds}\frac{d\theta}{ds} = \frac{1}{r}\frac{\partial\eta}{\partial\theta} \tag{7.51}$$

The left-hand side of equation 7.51 is

$$\frac{1}{r}\frac{d}{ds}\left(\eta r^2 \frac{d\theta}{ds}\right)$$

giving

$$\frac{d}{ds}\left(\eta r^2 \frac{d\theta}{ds}\right) = \frac{d\eta}{d\theta} \tag{7.52}$$

For a ray confined to a diametral plane

$$\frac{1}{r}\frac{dr}{d\theta} = \tan \alpha \qquad r\frac{d\theta}{ds} = \cos \alpha \qquad \frac{dr}{ds} = \sin \alpha$$

Then

$$\frac{d}{ds} \equiv \frac{d}{d\theta}\left(\frac{d\theta}{ds}\right) \equiv \frac{1}{r}\cos \alpha \frac{d}{d\theta} \tag{7.53}$$

and hence from equation 7.50

$$\frac{d}{ds}(\eta \sin \alpha) - \frac{\eta}{r}\cos^2 \alpha = 0$$

which leads to

$$\frac{\tan \alpha}{\eta} \frac{d\eta}{d\theta} = 1 - \frac{d\alpha}{d\theta} \qquad (7.54)$$

Equation 7.51 produces the same result.

Putting $r'/r = \tan \alpha = v$ we obtain

$$\frac{\eta'}{\eta} = \frac{1}{v}\left[1 - \frac{v'}{(1 + v^2)}\right] \qquad (7.55)$$

the prime referring to differentiation with respect to θ.

Expanding the second term in equation 7.55 by partial fractions and integrating, we obtain

$$\eta = \frac{C(1 + v^2)^{\frac{1}{2}}}{v} \exp\left(\int \frac{1}{v} d\theta\right); \qquad v = \frac{r'}{r} \qquad (7.56)$$

Thus given any one parameter pencil of rays, $r = r(a, \theta)$, the medium refractive index given by equation 7.56 is a solution, only if the parameter a disappears in the derivation. On the other hand, the definition of the ray trajectory for a specified angular refractive medium depends on the solution of equation 7.56 for $r(\theta)$ given $\eta(\theta)$. As can be seen, this is a very complicated problem.

It is important in what follows to remember that the θ coordinate is measured from the pole in the analysis, and thus the usual polar coordinate expressions for the curves we are to consider as in Appendix I, will be modified by the substitution of sine for cosine, and vice versa. The importance of this arises mainly in the effect it has on the sign of exponent in equation 7.56, thus inverting the solution.

The general rules of rays in a variable refractive medium also apply in this case. That is, the rays can never become orthogonal to the stratification of the medium without an infinite value of the refractive index occurring. In particular in this case, and contrary to expectations, circular rays with centre at the origin are impossible to achieve. This creates a severe limitation to those ray patterns that can be created by a realistic refractive medium, that is a medium with $\infty > \eta \geqslant 1$. Since the solutions are given in terms of $v = r'/r$, the ray patterns are form-invariant for scale changes in the radial coordinate. As a result, most of the examples will deal with spiral forms of rays, although for other forms finite-valued regions of the refractive medium can be delineated as boundary conditions in the solution of equation 7.56.

7.5.1 Spiral rays

With $v = \text{constant} = a$

$$r' = ar \qquad \text{and thus} \qquad r = e^{a\theta} \qquad (7.57a)$$

The rays are thus equiangular spirals and direct substitution into equation 7.56

120

provides the refractive index

$$\eta = A\mathrm{e}^{\theta/a} \tag{7.57b}$$

the constant A containing all the other constant factors.

7.5.2 Circular rays passing through the origin

With

$$r = a \sin \theta \qquad (0 \leqslant \theta \leqslant \pi) \tag{7.58a}$$

and therefore $r' = a \cos \theta$, we have $v = \cot \theta$. This gives

$$\eta = \frac{C(1 + \cot^2 \theta)^{\frac{1}{2}}}{\cot \theta} \exp\left(\int \tan \theta \mathrm{d}\theta\right) = C(\cos \theta)^{-2} \tag{7.58b}$$

This function becomes infinite on the axis $\theta = \pi/2$ where the rays are perpendicular to the stratification.

If the medium is terminated at a realistic value of the refractive index, at a value θ_0 say, then the rays will all be incident upon the surface $\theta = \theta_0$ at the same angle. Consequently they will all refract by the same amount into free space, giving a collimated pencil of parallel rays. The angle of emission can be chosen quite arbitrarily to create any required cone of rays, upon rotation of the system about the axis $\theta = 0$. This is illustrated in Figure 7.21. With horizontal rays as shown, full azimuthal coverage is obtained from the now conical angularly non-uniform lens. The singularity at the tip can be removed by cutting the lens with a plane section, AA in Figure 7.21, leaving a circular aperture in which a line at infinity is imaged into a horizontal radial line. This is the effect of the camera obscura.

7.5.3 The sinusoidal spirals

As shown in Appendix I these have the general equation

$$r^n = a^n \sin n\theta \tag{7.59a}$$

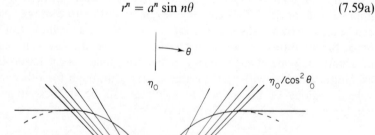

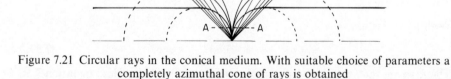

Figure 7.21 Circular rays in the conical medium. With suitable choice of parameters a completely azimuthal cone of rays is obtained

and $v = r'/r = \cot n\theta$, and hence from equation 7.56

$$\eta = A(\cos n\theta)^{-(1+n)/n} \tag{7.59b}$$

The straight line $n = -1$ thus requires a uniform refractive medium. Similar results can be obtained with the multi-folium curves

$$r = b \sin m\theta \tag{7.60a}$$

Then $v = m \cot m$, and therefore

$$\eta = \frac{C(\sin^2 m\theta + m^2 \cos^2 m\theta)^{\frac{1}{2}}}{m(\cos m\theta)^{1+1/m^2}} \tag{7.60b}$$

7.5.4 The Archimedean spirals

The general relation is (Appendix I)

$$r^m = a^m\theta \tag{7.61a}$$

Differentiating gives $v = 1/m\theta$ and consequently

$$\eta = Ae^{m\theta^2/2}(1 + m^2\theta^2)^{\frac{1}{2}} \tag{7.61b}$$

Spiral curves that approach the circular inevitably give rise to refractive-index laws that rapidly become infinitely large. For example, the curve $r = a \tanh (\theta/2)$ requires the refractive index $\eta = A \cosh \theta e^{\cosh\theta}$ which is excessive even within the permitted range of $0 \leqslant \theta \leqslant 2\pi$.

Numerous other curves can be illustrated in this manner, including the curves $r = a \sin^n \theta$ and the cardioids. For final illustration we show those curves which tend to and eventually pass beyond parallelism with the stratification. The designs are suitable for beam expanders or spot size reduction.

7.5.5 Parabolic rays

The appropriate polar equation for the parabola is

$$r = 4a \cos \theta/\sin^2 \theta \tag{7.62a}$$

and differentiation gives $v = -(\sin^2 \theta + 2 \cos^2 \theta)/\sin \theta \cos \theta$. The exponent in equation 7.56 is thus

$$-\int \frac{\sin \theta \cos \theta}{\sin^2 \theta + 2 \cos^2 \theta} \, d\theta = \tfrac{1}{6}\log (\sin^2 \theta + 2\cos^2 \theta)$$

and hence

$$\eta = \frac{C(1 + v^2)^{\frac{1}{2}}}{v} (\sin^2 \theta + 2\cos^2 \theta)^{\frac{1}{6}} \tag{7.62b}$$

with v given above.

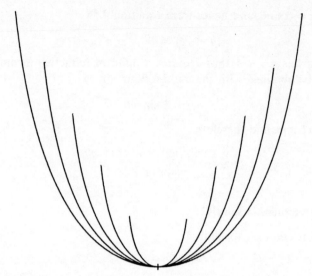

Figure 7.22 Rays in the conical medium with refractive law of equation 7.63b—the kappa curves $r = a \cot \theta$

This is one of many instances that occur in which both r'/r and r/r' can be derived in terms of elementary integrals.

7.5.6 The kappa curves (Figure 7.22)

These are similar to parabolic curves with polar equation (Appendix I)

$$r = a \cot \theta \tag{7.63a}$$

Hence (Figure 7.23)

$$v = -2/\sin 2\theta \qquad \eta = \frac{C}{2}(4 + \sin^2 2\theta)^{\frac{1}{2}} e^{\cos 2\theta/4} \tag{7.63b}$$

Since the entire analysis has been shown to be derived in terms of the variable $v = r'/r$, in an inversion $r \to 1/R$, all angles remain invariant and

$$R'/R = -r'/r = -v$$

The negative sign in the denominator can be incorporated in the constant of integration C, and thus the sole effect of the inversion is to invert the exponential factor in the solution.

7.5.7 The epi spirals

These are the inverses of the multi-folium curves of equation 7.60a, and their equations are thus

$$r = a/\sin m\theta \qquad m = 1, 2, 3 \tag{7.64a}$$

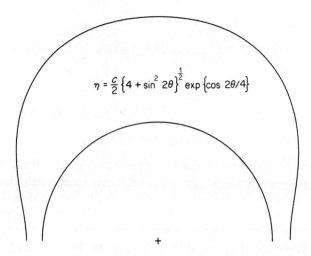

$$\eta = \frac{C}{2}\left\{4 + \sin^2 2\theta\right\}^{\frac{1}{2}} \exp\left\{\cos 2\theta/4\right\}$$

Figure 7.23 Refractive index variation for kappa curve rays

Hence $v = -m \cot m\theta$, the negative of that in equations 7.60, and results in the refractive index

$$\eta = \frac{C(\sin^2 m\theta + m^2 \cos^2 m\theta)^{\frac{1}{2}}}{-m} (\cos m\theta)^{(1/m)-1} \qquad (7.64b)$$

giving the same m-fold singularity conditions at $m\theta = (2n + 1)\pi/2$.

8 Geodesics, Rays and Trajectories

8.1 GEODESIC LENSES

There are numerous ways in which a microwave or an optical field can be confined to a surface layer. In the case of electromagnetic radiation, a pair of infinite plane conductors closely spaced and excited by the TEM mode, that is by a totally transverse electric field vector, confines the field to the planar space between the conductors. A point source of radiation between the planes would radiate isotropically in the space between the conductors and the refractive index would be uniform and unity. If, as shown in Figure 8.1, the two conductors were to be formed into curved surfaces remaining 'parallel' in the sense of the geometry of surfaces, the radiation would proceed along rays which are the geodesics of a surface intermediate between the two conducting surfaces (and usually taken to be midway between the two).

Two infinite plane conductors excited by the dominant H_{01} mode still contains an isotropic space but with a refractive index less than unity and dependent upon the spacing between them. The apparent index for planes separated by a distance h is given by

$$\eta = [1 - \lambda^2/(2h)^2]^{\frac{1}{2}} \qquad (8.1)$$

λ being the wavelength of the excitation. In coordinates of η and $\lambda/2h$ this is a circle of unity radius. The space between the conductors could be filled if required with a lossless dielectric. With this mode of propagation a variable medium can be constructed by making the separation h a function of position. This is ideally suited to the practical realization of the angular medium described in the previous chapter. We take, for example, the rays in the form of kappa curves $r = a \cot \theta$ shown in Figure 7.22; after a short distance the rays become virtually parallel, and thus a horn with curved faces that produces this ray configuration would be corrected for the phase curvature that occurs in the ordinary pyramidal horn. The profile is then obtained by substituting the refractive-index law of equation 7.63b into equation 8.1 and solving for the separation h as a function of θ. That is

$$\left(1 - \frac{\lambda^2}{4h^2}\right)^{\frac{1}{2}} = \frac{C}{2}(\sin^2 2\theta + 4)^{\frac{1}{2}}e^{\cos 2\theta/4}$$

and hence

$$h = \frac{\lambda}{2\left[1 - \frac{C^2}{4}(\sin^2 2\theta + 4)e^{\cos 2\theta/2}\right]^{\frac{1}{2}}} \qquad (8.2)$$

an illustration of which is shown in Figure 8.2.

124

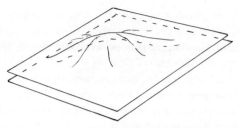

Figure 8.1 Geodesic surface confined between two parallel curved conducting plates

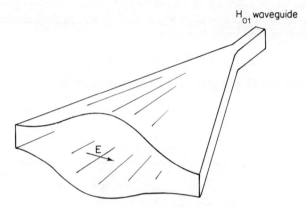

Figure 8.2 Non-uniform angular medium waveguide analogue

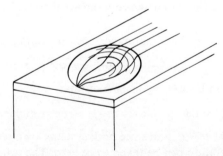

Figure 8.3 Geodesic lens confined to a thin dielectric layer

Light, too, can be guided isotropically by a thin layer of uniform refractive material and confined, as in the optical fibre, by total internal reflection. In the previous chapters we have discussed the effects of making such a layer non-uniform in its medium properties, or equivalently of non-uniform thickness. However, we can retain uniformity both of material and thickness and obtain ray divergence by distorting the surface from the plane (Figure 8.3). The rays then follow the geodesics of these surfaces. Naturally for mechanical reasons such layers are supported by a substrate of lower refractive index, and hence the surface curvatures can be machined into such substrates.

126

There is a one-to-one relation between the spherical non-uniform lens designs of the previous chapter and the circularly symmetrical surface, a surface of revolution, based upon the comparison between the rays as geodesics in the two systems. The resulting surfaces are collectively known as geodesic lenses and their optical properties are identical to the non-uniform spherical lenses (in a diametral cross-section). Such designs have long been applied to microwave antennas.[39]

A fact of some importance that emerges from the analysis is that a refractive medium with a singularity in the refractive index can still be represented by a real and practicable geodesic surface. This is illustrated by the simple case of the equivalent refractive index for the geodesics on a right circular cone.

For a circularly symmetrical geodesic surface the optical increment of the geodesic path is, as shown in Figure 8.4

$$\eta^2 ds^2 = d\sigma^2 + \rho^2 d\phi^2 \tag{8.2}$$

since the refractive index is unity for TEM propagation.

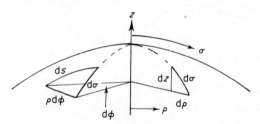

Figure 8.4 Geodesics on a surface of revolution

The optical distance in the nonhomogeneous refracting lenses is

$$\eta^2 ds^2 = \eta^2(r)(dr^2 + r^2 d\theta^2) \tag{8.3}$$

For equivalence it can be seen by inspection that

$$d\theta = d\phi \qquad \rho(\sigma) = r\eta(r) \qquad d\sigma = \eta(r)dr \tag{8.4}$$

From the first of these $\theta = \phi$, and since both systems are circularly symmetrical the constant of integration can be taken to be zero. The transformation is then between η as a function of r and σ as a function of ρ.

With this transformation it is possible to convert most of the known lens solutions of equation 7.6, together with their known ray properties, into geodesic surface profiles with the same ray properties. When this is done, as shown in the Table, all the $\sigma(\rho)$ formulations can be seen to be included in a completely general formulation[40]

$$\sigma = A\rho + B \sin^{-1} C\rho \tag{8.5}$$

characterized by the three apparently arbitrary parameters A, B and C.

Consequently we use this general formulation to derive, by the reverse of the procedure, a general formulation for non-uniform refractive lenses using the same

Generalized law of refractive index $Cr^{2/BC} - 2r^{1/BC}(\eta r)^{(A/BC)-1} + C(\eta r)^{2A/BC} = 0$

Lens form	$\eta(r)$	A	B	C	Reference
Maxwell's fish-eye	$\dfrac{2}{1+r^2}$	0	1	1	Equation 7.9
Generalized fish-eye (resonator ring)	$\dfrac{2r^{\lambda-1}}{1+r^{2\lambda}}$	0	$1/\lambda$	1	Equation 7.11
Luneburg	$(2-r^2)^{\frac{1}{2}}$	$\frac{1}{2}$	$\frac{1}{2}$	1	Equation 7.14
Eaton (transform of fish-eye $\eta \leftrightarrow r$)	$(2/r-1)^{\frac{1}{2}}$	1	1	1	Equation 7.18
Guttman	$(1+f^2-r^2)^{\frac{1}{2}}/f$	$\frac{1}{2}$	$\dfrac{1+f^2}{4f}$	$\dfrac{2f}{1+f^2}$	Equation 7.16
Toraldo	$\eta^2 r^2 = \eta^{1/p}(2-\eta^{1/p})$	1	p	1	Equation 7.20
Transformed Toraldo ($\eta \leftrightarrow r$)	$(2r^{1/p}-r^{2/p})^{\frac{1}{2}}/r$	p	p	1	Equation 7.28
Cone (semivertex angle ψ: $A=\operatorname{cosec}\psi$)	$r^{(1/A-1)}$	A	0	0	Equations 8.20 and 8.11
Plane	1	1	0	0	
Lens with hyperbolic rays (including complex values $K>2$)	$[2K+(K^2-2K+r^2)^{\frac{1}{2}}]^{\frac{1}{2}}$	$\frac{1}{2}$	$\dfrac{K}{2(2K-K^2)^{\frac{1}{2}}}$	$\dfrac{(2K-K^2)^{\frac{1}{2}}}{K}$	Equation 7.21
Beam divider (point source) through angles $\pm p\pi/2$	Equation 7.14	$\frac{1}{2}$	$-\dfrac{(p+1)}{2}$	1	
Beam divider (parallel ray source) through angles $\pm q\pi/2$	Equation 7.17	1	$\dfrac{2-q}{2}$	1	

three parameters.[41] Thus from equations 8.5 and 8.6

$$d\sigma = Ad\rho + \frac{BC}{(1 + C^2\rho^2)^{\frac{1}{2}}} d\rho = \eta dr = \frac{\rho}{r} dr$$

$$\frac{dr}{r} = \frac{d\rho}{\rho} \left[A + \frac{BC}{(1 - C^2\rho^2)^{\frac{1}{2}}} \right]$$

Putting $C\rho = \sin \gamma$, we have

$$\log r = A \log \rho + BC \log \tan (\gamma/2)$$

or

$$r = \rho^A \left[\frac{1 - (1 - C^2\rho^2)^{\frac{1}{2}}}{1 + (1 - C^2\rho^2)^{\frac{1}{2}}} \right]^{BC/2} \tag{8.6}$$

giving, since $\rho = \eta r$

$$C\eta r[r^{2/BC} + (\eta r)^{(2A/BC)}] = 2r^{1/BC}(\eta r)^{A/BC}$$

or

$$Cr^{2/BC} - 2r^{1/BC}(\eta r)^{(A/BC)-1} + C(\eta r)^{2A/BC} = 0 \tag{8.7}$$

It can readily be confirmed that the appropriate values of the parameters result in the refractive-index laws shown in the Table.

Some points of interest arise immediately from this result. First, a change in sign of the parameter B leaves the result unchanged. This can only mean that there are *two* geodesic surfaces (in some instances) lying on each side of the cone $\sigma = A\rho$, which have the same ray properties.

Secondly, the transformation $\eta \to r$ and $r \to \eta$, transform pairs of lenses with different properties as shown in the Table. This transformation, also the result of a Legendre transformation (equation 7.4), applied to equations 8.6 or 8.7 shows it to remain form-invariant if A is replaced by $1 - A$. This simple rule is thus the equivalent of a Legendre transformation.

In the form given by equation 8.5 it is a simple matter to establish a condition among the parameters that will give the geodesic surface the highly desirable property of becoming tangentially flat at the periphery. There will then be no edge giving ray diffraction effects. This requires $d\sigma/d\rho = 1$ when $\rho = 1$, or

$$(1 - C^2)(1 - A)^2 = B^2 C^2 \tag{8.8}$$

In the same way a geodesic lens that is smooth and flat about the axis has $d\sigma/d\rho = 1$ at $\rho = 0$, or

$$A + BC = 1 \tag{8.9}$$

These conditions obviously cannot be met simultaneously as the combined result gives $C = 0$, which corresponds once more to a cone $\sigma = A\rho$ which clearly meets neither criterion.

In order to use the ray-trace equation it is necessary to derive η explicitly as a function of r from the relations in equations 8.6 or 8.7. Only particular values of the parameters make this possible and these include the complete range of known refractive devices. Obvious conditions for an explicit solution to equation 8.7 are

   (i) $A = 0$
   (ii) $A = BC$
   (iii) $BC = 0$
   (iv) $A = -BC/3$

to which can be added the result of a change of sign in B. The other transformation $A \rightarrow 1 - A$ does not necessarily transform an explicit solution into another explicit solution, as can be seen from the Table when applied to the Toraldo lens.

   With condition (i) we obtain

$$\eta = \frac{2}{C} \frac{r^{(1/BC)-1}}{1 + r^{2/BC}}$$

the generalized fish-eye solutions. In particular, if $BC = 1$ we have the original fish-eye for which the geodesic analogue is $\sigma = \sin \rho$, which is the equation of a circle. On rotation this is therefore the sphere, the stereographic projection of the great circles being the coaxal system of rays shown in Figure 7.3. This was the method first used by Luneburg[25] to derive this result.

   With condition (ii)

$$A = \pm BC \qquad \eta^2 = \frac{2}{r^2} (r^{1/BC} - Cr^{2/BC})$$

the Luneburg solutions (cf. equations 7.28) which are non-singular if $BC \geqslant 2$.

   Condition (iv) is highly unusual. With $A = -BC/3$ we obtain from equation 8.7

$$2r^{-(1/3A)}(\eta r)^{-4/3} - C(\eta r)^{-2/3} - Cr^{-(2/3A)} = 0$$

This result has not been investigated further.

   Condition (iii), referred to equation 8.7, results in the refractive-index law

$$\eta = r^{\nu - 1} \tag{8.10}$$

and the geodesic surface

$$\sigma = \frac{1}{\nu} \tag{8.11}$$

This is a right circular cone of semi-vertex angle ψ, where $\nu = \sin \psi$. Putting $A = 1/\nu$, then equation 8.10 represents a refractive index with an infinite singularity at the origin, corresponding to the real geodesic surface, the cone. Substituting $\eta = r^{(1/A)-1}$ into the ray equation (equation 7.7) we have

$$\phi - \phi_0 = \int_{r_0}^{r} \frac{K dr}{r(r^{2/A} - K^2)^{\frac{1}{2}}} \tag{8.12}$$

the sign depending on the parts of the ray for which r increases or decreases with

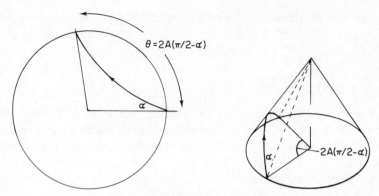

Figure 8.5 (a) Ray trajectory in the non-uniform medium analogous to the cone
(b) Ray trajectory as the geodesic on a cone

ϕ. This changes at the point r_m, where the ray is at its closest to the centre and consequently perpendicular to the radius vector. The ray starts from the periphery, that is $r_0 = 1$, making an angle α with the diameter (Figure 8.5(a)). Consequently $K = \sin \alpha$, and from the invariance of $\eta r \sin \psi = K$

$$\eta_m r_m = K \qquad r_m = (\sin \alpha)^A$$

Substitution of these values into equation (8.12) gives

$$\phi - \phi_0 = 2 \int_1^{r_m} \frac{\sin \alpha \, dr}{r(r^{2/A} - \sin^2 \alpha)^{\frac{1}{2}}}; \qquad r_m = \sin \alpha^A$$

The substitution $r^{1/A} = \sin \alpha \operatorname{cosec} \beta$ results in

$$\phi - \phi_0 = A \operatorname{cosec}^{-1} \left(\frac{r^{1/A}}{\sin \alpha} \right) \Big|_1^{(\sin \alpha)^A}$$

Taking ϕ_0 to be the zero angle of reference

$$\phi = 2A \left(\frac{\pi}{2} - \alpha \right) \tag{8.13}$$

the cone, being a developable surface, has geodesics which are straight lines on the developed surface. Figure 8.5(b) shows that such a geodesic will have the result given in equation 8.13. As a consequence of this result a refractive lens constructed of n uniform concentric shells with refractive-index laws $r^{1/A_n - 1}$ has a geodesic analogue consisting of n continuous cone frusta with cone angles $A_n = \operatorname{cosec} \psi_n$.

It should be noted that the reverse procedure, that is specification *a priori* of the ray pattern, even where it gives rise to implicit functions of r and $\eta(r)$, have all been reducible to the form of equation 8.7, for example that of equation 7.14.

The results given so far are in the form of a (σ, ρ) equation. For machining purposes this has to be converted to a (z, ρ) formulation. The methods inevitably require a computational procedure.

Combining the two results

$$d\sigma^2 = dz^2 + d\rho^2$$

and

$$d\sigma = \left(A + \frac{BC}{1 - C^2\rho^2}\right)d\rho$$

then

$$dz^2 = d\rho^2\left[A^2 + \frac{B^2C^2}{1 - C^2\rho^2} + \frac{2ABC}{(1 - C^2\rho^2)^{\frac{1}{2}}} - 1\right] \qquad (8.14)$$

This formula can be put into a finite-difference form and computed directly for incremental steps $\Delta\rho$. A check on the accuracy can be obtained by inserting the conditions for the Maxwell fish-eye and comparing the result with the true circle that should be obtained (no. 1 of the Table). On substituting, as before, $C\rho = \sin\gamma$, we obtain

$$z = \frac{1}{C}\int_0^{\sin^{-1}C\rho}[(A^2 - 1)\cos^2\gamma + 2ABC\cos\gamma + B^2C^2]^{\frac{1}{2}}d\gamma \qquad (8.15)$$

The profile computed for the beam divider given by equation 7.17 is shown in Figure 8.6.

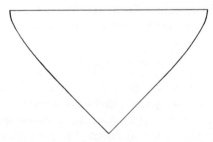

Figure 8.6 The geodesic surface for the beam divider of Figure 7.10

In all these lenses, as in the non-uniform lenses, the central ray is a geodesic in its own right. In the devices which deflect the main body of rays, the central ray would require independent treatment to be deflected in a manner compatible with the ray pattern. This could take the form of a simple reflecting prism or wedge placed in the medium at its centre.

For those devices not meeting the criterion of equation 8.8, the edge discontinuity could cause scattering and loss effects. A 'rounded edge' procedure has been outlined[42] for this condition, and its incorporation into the lenses included here has to be carried out early in the design to prevent the creation of additional aberrations.

One method would be to determine the geodesic relation describing such a 'rounded edge' and converting this to a non-uniform refractive layer surrounding the analogous non-uniform lens. Performing a ray-trace operation on this combination by means of the standard integral (equation 7.7) will give the magnitude of the aberrations created by the edge-rounding process. Alternatively a 45° chamfered edge could be used. This would be the frustum of a cone and its equivalent refractive index of the shell is given by equation 8.10.

The effect of the inversion on this lens formula is to change the sign of A in the general relation of equation 8.5. This has been obtained from the tabulated examples only and not generally proved. In most cases this still leaves a positive σ value and a realistic surface. The result can, however, leave a real refractive index attached to an unrealistic geodesic surface. This can again be seen in the case of the cone.

The refractive index for a cone of semi-vertex angle ψ with $A = \operatorname{cosec} \psi$ is

$$\eta = r^{(1/A)-1}$$

This inverts to $\bar{\eta} = r^{-(1/A)-1}$ valid except at $r = 0$, but the real cone $\sigma = A\rho$ becomes $\sigma = -A\rho$, which is unrealistic.

8.2 RAYS AND TRAJECTORIES

The optical ray as the path of least action of an action function has naturally a lot in common with other trajectories based on the same principle. In mechanics the principle of Fermat and Maupertuis was incorporated by Hamilton in its application to a characteristic function. The result is an identical set of Hamilton's equations for the trajectories of particles as given by equation 5.8, namely

$$(\nabla W)^2 = 2m(E - V) \tag{8.16}$$

Here W is the characteristic function of the canonical coordinates, p the generalized momenta and q the generalized coordinates,[43] where for a holonomic system (equivalent in optics to a non-dispersive medium) the Hamiltonian is identical to the total energy. The principle of least action for the dynamical system is then

$$\delta \int [2m(E - V)]^{\frac{1}{2}} \mathrm{d}s = 0 \tag{8.17}$$

which by analogy associates $2m(E - V)$ with η^2 as do equations 5.8 and 8.16.

It is of interest to observe that the three major fields covered both by Newton and Hamilton were optics, mechanics and algebra, the last of these being in both cases the required mathematics to apply to the former. Hamilton's mechanics is basically the applications of the action principle to the characteristic function W and his great contribution to optics was the derivation of similar characteristic functions suitable for that subject. This analogy is well known, but in view of the fact that there has been derived in this work a transformation applicable to the optics field, it is worth enquiring where a similar transformation theory arises in the field of particle trajectories.

A somewhat similar situation arises in considering the geometrical optics rays as first-order solutions of the electromagnetic field equations. In that case a transformation such as Damien's inversion in Chapter 3 should have a concomitant in electromagnetic field theory.

We shall summarize here some of the similarities and contradictions that arise when one compares these physical systems with the transformation theorems in mind.

Recently Buchdahl[44] has compared the equivalence of the problems of the rays in Maxwell's fish-eye and Kepler orbits, the archetypal solutions in each discipline. Two constant vectors are known in the dynamical system, the angular momentum and the Runge–Lenz vector. Bouguer's theorem for a ray in a spherical medium is shown to be the analogue of the constancy of the angular momentum and a second optical vector is analogous to the Runge–Lenz vector. However, the connection is between a circular ray and an elliptical orbit. The elliptical orbit in real space is a circular trajectory in the phase space of generalized coordinates. Alternatively the circular rays of the fish-eye can be converted to the proper Kepler orbits, the rays in the Eaton lens (see Figure 7.11) by a Legendre transformation. Buchdahl continues the discussion with a study of the invariance of this optical system under conformal transformations. The inversion is a special form of such a conformal transformation.

Similarly a discussion of the classical deflection function, Rutherford scattering, gives a new deflection formula which is the ray integral of equation 7.6. Luneburg had already shown that the refractive-index law

$$\eta^2 = C + \frac{1}{r}$$

gives rays which are hyperbolic if $C > 0$, parabolic if $C = 0$ and elliptical (the Eaton lens) if $C < 0$.[25] There is, too, a transformation of the potential function $V(r)$ in equation 8.17 which gives the trajectories of particles in the transformed potential exactly as we have directly obtained the rays in the inverted refractive medium. This dynamical transformation[45] is an inversion acting on the phase-space of the mechanical motion.

Mechanical problems also have an analogous geodesic surface of revolution,[46] and many transformations of the mechanical solution were obtained from the permissible transformations of this surface. These included stereographic projection, conformal, geodesic or isometric mappings. Such transformations could be applied to the geodesic surface of revolution in the optical analogue.

The natural families of trajectories, of which the optical ray is one class, consists of all problems governed by the extremum of an action integral, and it has been proved that the only point transformations which convert every natural family of trajectories into a natural family are those belonging to the conformal group to which, of course, the inversions belong.

The group of conformal transformations in four dimensions can be built up of transformations by inversions in the hyperspheres of that space.[47] As such they are applicable to electromagnetic field problems. There are two fundamental transformations under which the electromagnetic field equations are covariant—first the Lorentz transformation, which can be generalized by the addition of translations and is then represented by the Poincaré group, and secondly the inversions which can be generalized in the same way to give the Möbius transformation represented by the conformal group. Both can be extended to include complex values of the group parameters.[48]

With the inclusion of a time dimension, the inversions are accompanied by a contraction (or expansion).[11] In this case, as opposed to the Lorentz case, the condition $t = 0$ can be applied simultaneously in both frames, leaving a spatial inversion. This condition, $T = t = 0$, could also indicate the existence of the zero-distance phase front. In the contracting frame it is interesting to find that the constituent relations undergo the identical transformation as in the Lorentz transformation. The two transformations have the same fundamental basis—they are the only transformations which transform a spherical wave into a spherical wave. In the Lorentz transformation, this invariance of the isotropy of space is sufficient to *prove*, that is not require, the constancy of the velocity of light.

For completeness, the field transformations of Cunningham for the inversion are

$$\mathbf{E}' = d^{-4}(r^2 - c^2t^2)[(r^2 - c^2t^2)\mathbf{E} + 2\mathbf{r} \times (\mathbf{r} \times \mathbf{E}) - 2ct(\mathbf{r} \times \mathbf{H})]$$

$$\mathbf{H}' = d^{-4}(r^2 - c^2t^2)[-(r^2 - c^2t^2)\mathbf{H} - 2\mathbf{r} \times (\mathbf{r} \times \mathbf{H}) - 2ct(\mathbf{r} \times \mathbf{E})] \qquad (8.18)$$

where d is the radius of inversion and $r^2 = x^2 + y^2 + z^2$. It can be seen that, for $t = 0$, ray directions $(\mathbf{E}' \times \mathbf{H}').\hat{\mathbf{r}} = (\mathbf{E} \times \mathbf{H}).\hat{\mathbf{r}}$ since in an inversion $\hat{\mathbf{r}}' = \hat{\mathbf{r}}$.

Subsequently we find that this inversion does not agree in two particulars with the two inversions we have demonstrated in the text. It is explicitly stated in the reference that a reflector is transformed into a reflector and *not*, therefore, into the zero-distance phase front, as required by Damien's theorem. Furthermore, with zero contraction the refractive index undergoes no transformation at all, while the ray directions are inverted.

Thus there are many avenues which are opened up as a result of the optical inversion theorems that have been demonstrated in the foregoing. Some discussion of the possible methods to approach these problems, in a way that will unify the optical with both the electromagnetic and the dynamical theories, is taking place.[49] The solution is for a future volume on this subject.

The conclusion remains, however, that there is considerably more to the geometry of geometrical optics than meets the eye. This is borne out by many authors who illustrate the central position geometrical optics plays, as the fundamental first-order solution of many physical processes. To quote one such author,[2] on finding that the unit triad of vectors at any point on a ray multiplied by the refractive index at that point obeyed Maxwell-like equations:

'One certain implication is that the geometrical component in physical optics is greater than anyone thought and that the usual methods for extracting geometrical optics from Maxwell's equations may be too restrictive'.

Appendix I
Curves and Their Formulae[6]

(a) *Astroid*—see *hypocycloid*

(b) *Cardioid*
    (i) $(x^2 + y^2 - 2ax)^2 = 4a^2(x^2 + y^2)$
    (ii) $r = 2a(1 + \cos \theta)$
    (iii) $x = 2a \cos t(1 + \cos t)$
         $y = 2a \sin t(1 + \cos t)$ $-\pi \leqslant t \leqslant \pi$

(c) *Catenary*
    (i) $y = a \cosh (x/a)$

(d) *Cayley's sextic*
    (i) $4(x^2 + y^2 - ax)^3 = 27a^2(x^2 + y^2)^2$
    (ii) $r = a \cos^3 (\theta/3)$

(e) *Cissoid* (Diocles)
    (i) $y^2(a - x) = x^3$
    (ii) $r = a \sin \theta \tan \theta$
    (iii) $x = a \sin^2 t;$ $y = a \tan t \sin^2 t$ $-\pi < t < \pi$

(f) *Circle*
    (i) $(x - h)^2 + (y - k)^2 = a^2$
    (ii) $r = a$
    (iii) $x = a \cos t;$ $y = a \sin t$ $-\pi \leqslant t \leqslant \pi$

(g.1) *Cubic* (Tschirnhausen)
    (i) $27ay^2 = x^2(x + 9a)$
    (ii) $r \cos^3 (\theta/3) = a$
    (iii) $x = 3a(t^2 - 3);$ $y = at(t^2 - 3)$ $-\infty < t < \infty$

(g.2) *Semi-cubical parabola*
    (i) $27ay^2 = 4x^3$
    (ii) $4r = 27a \sin^2 \theta \sec^3 \theta$
    (iii) $x = 3at^2;$ $y = 2at^3$ $-\infty < t < \infty$

(h) *Cycloid*
    (iii) $x = at - h \sin t;$ $y = a - h \cos t$ $-\infty < t < \infty$

(i) *Ellipse*

    (i) $x^2/a^2 + y^2/b^2 = 1$

    (ii) $r^2 = b^2/(1 - e^2 \cos^2 \theta);$ $ar = b^2/(1 + e \cos \theta)$

    (iii) $x = a \cos t;$ $y = b \sin t$ $-\pi \leqslant t \leqslant \pi$

        $b^2 = a^2(1 - e^2)$

        $e \equiv$ eccentricity

(j) *Epicycloid*

    (iii) $x = m \cos t - b \cos (mt/b)$

        $y = m \sin t - b \sin (mt/b)$ $-\pi \leqslant t \leqslant \pi$

        $m = a + b$

     $a = b$ see *cardioid*; $a = 2b$ see *nephroid*

(k) *Hyperbola*

    (i) $x^2/a^2 - y^2/b^2 = 1;$ $b^2 = a^2(e^2 - 1)$

        $xy = c^2$

    (ii) $r^2 = b^2/(e^2 \cos^2 \theta - 1)$ (equilateral if $a = b$)

        $ar = b^2/(1 + e \cos \theta)$

        $2r^2 \sin^2 \theta = a^2 e^2$

    (iii) $x = a \sec t;$ $y = b \tan t$ $-\pi \leqslant t \leqslant \pi$

        $x = ct;$ $y = c/t$

        $e \equiv$ eccentricity

(l.1) *Hypocycloid*

    (iii) $x = n \cos t + b \cos (nt/b)$

        $y = n \sin t - b \sin (nt/b)$ $-\pi \leqslant t \leqslant \pi$

        $n = a - b$

     $b = 0$ see *circle*

(l.2) *Astroid* $(a = 4b)$

    (i) $x^{\frac{2}{3}} + y^{\frac{2}{3}} = a^{\frac{2}{3}}$

    (iii) $x = 4a \cos^3 t;$ $y = 4a \sin^3 t$

(m) *Lemniscate* (Bernoulli)

    (i) $(x^2 + y^2)^2 = a^2(x^2 - y^2)$

    (ii) $r^2 = a^2 \cos 2\theta$

    (iii) $x = a \cos t/(1 + \sin^2 t)$

        $y = a \sin t \cos t/(1 + \sin^2 t)$ $-\pi \leqslant t \leqslant \pi$

(n) *Limaçon* (Pascal)

    (i) $(x^2 + y^2 - 2ax)^2 = b^2(x^2 + y^2)$

    (ii) $b^2 r^2 = (r^2 - 2ar \cos \theta)^2$

    (iii) $x = \cos t(2a \cos t + b)$

        $y = \sin t(2a \cos t + b)$ $-\pi \leqslant t \leqslant \pi$

(o) *Nephroid* (Huygens, Figure I.1)

    (i) $(x^2 + y^2 - 4a^2)^3 = 108a^4 y^2$

    (ii) $(r/2a)^{\frac{2}{3}} = (\sin \theta/2)^{\frac{2}{3}} + (\cos \theta/2)^{\frac{2}{3}}$

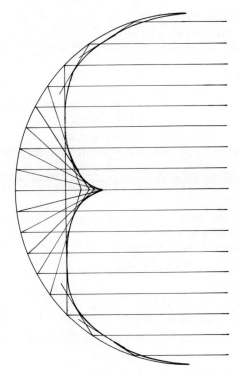

Figure I.1 The nephroid

(iii) $x = a(3 \cos t - \cos 3t)$
$\quad\;\; y = a(3 \sin t - \sin 3t) \qquad -\pi \leqslant t \leqslant \pi$

(p.1) *Oval* (Descartes)
$\quad$ (i) $[b(x^2 + y^2 + 1) - 2cx]^2 = 2c(x^2 + y^2 + 1) - 4bx - 1$

(p.2) *Oval* (Cassini)
$\quad$ (i) $(x^2 + y^2 + a^2)^2 = b^2 + 4a^2x^2$
$\quad$ (ii) $r^4 - 2a^2r^2 \cos 2\theta = b^4 - a^4$

(q) *Parabola*
$\quad$ (i) $y^2 = 4ax$
$\quad$ (ii) $r = 2a/(1 - \cos \theta)$
$\quad$ (iii) $x = at^2; \qquad y = 2at \qquad -\infty < t < \infty$

(r.1) *Spiral* (Archimedes)
$\quad$ (ii) $r^m = a^m\theta$
$\quad$ $m = 1 \qquad$ Archimedean
$\quad$ $m = 2 \qquad$ Fermat
$\quad$ $m = -1$ Hyperbolic
$\quad$ $m = -2$ Lituus

(r.2) *Logarithmic* or *equiangular spiral*
    (i) $x^2 + y^2 = \exp\{2a \tan^{-1}(y/x)\}$
    (ii) $r = \exp(a\theta)$

(r.3) *Sinusoidal spirals*
    (ii) $r^n = a^n \cos n\theta$

$n = \frac{1}{3}$ Cayley's sextic	$n = -\frac{1}{3}$ Tschirnhausen's cubic
$n = \frac{1}{2}$ Cardioid	$n = -\frac{1}{2}$ Parabola
$n = 1$ Circle	$n = -1$ Straight line
$n = 2$ Lemniscate	$n = -2$ Equilateral hyperbola

(s)   *Tschirnhausen's cubic*—see *cubic*

(t)   *Tractrix* (Huygens)
    (i) $x = \pm a \cosh^{-1}(a/y) - (a^2 - y^2)^{\frac{1}{2}}$
    (ii) $r = a \tan \theta$ (origin at cusp)
    (iii) $x = a \ln(\sec t + \tan t) - a \sin t$
        $y = a \cos t \qquad -\pi/2 \leqslant t \leqslant \pi/2$

(u)   *Trisectrix* (Maclaurin)
    (i) $y^2(a - x) = x^2(x + 3a)$
    (ii) $r = a \sec \theta - 4a \cos \theta$
    (iii) $x = a(t^2 - 3)/(t^2 + 1); \qquad y = at(t^2 - 3)/(t^2 + 1) \qquad -\infty < t < \infty$

(v)   *Kappa curve*
    (i) $(x^2 + y^2)y^2 = a^2 x^2$
    (ii) $r = a \cot \theta$
    (iii) $x = a \cos t \cot t; \qquad y = a \cos t \qquad 0 < t < 2\pi$

(w) *Right strophoid*
    i) $y^2(a - x) = x^2(a + x)$
    ii) $r = a(\sec \theta - 2 \cos \theta)$
    iii) $x = a(1 - 2\cos^2 t)$
        $y = a \tan t(1 - 2\cos^2 t) \qquad -\pi \leqslant t \leqslant \pi$

Appendix II
Derived Curves[6]

II.1 CURVES AND THEIR INVERSES

Curve ←	Centre of inversion	→	Inverse
straight line	point not on line		circle
circle	point on circumference		straight line
	point not on circumference		circle
parabola	focus	cusp	cardioid
	vertex	cusp	cissoid
central conic	focus	pole	limaçon
	centre	centre	oval, figure of eight
rectangular	centre	centre	lemniscate
hyperbola	vertex	node	right strophoid
right strophoid	pole		same strophoid
sinusoidal spiral	pole		sinusoidal spiral
logarithmic spiral	pole		logarithmic spiral
Archimedean spiral	pole		Archimedean spiral

II.2 EVOLUTES AND INVOLUTES

Involute	Evolute
parabola	semi-cubical parabola
limaçon of Pascal	catacaustic of the circle
cardioid, parameter a	cardioid, parameter $a/3$
epicycloid	epicycloid
nephroid, parameter a	nephroid, parameter $a/2$
hypocycloid	hypocycloid
logarithmic spiral	equal logarithmic spiral
cycloid	equal cycloid
tractrix	catenary

II.3 CATACAUSTICS

Curve	Source	Catacaustic
circle	on circumference	cardioid
	not on circumference	limaçon of Pascal
	parallel rays	nephroid
parabola	axial parallel rays	focal point
	rays not parallel to axis	Tschirnhausen's cubic[13]
cardioid	cusp	nephroid
logarithmic spiral	pole	equal spiral
cycloidal arch	rays perpendicular to axis	two equal cycloidal arches
$y = \ln x$	rays parallel to axis	catenary
Tschirnhausen's cubic	focus	semi-cubical parabola
cissoid of Diocles	focus	cardioid
quadrifolium	centre	astroid

II.4 PEDAL CURVES

Curve	Pole	Pedal
line	any point	point
circle	any point	limaçon of Pascal
	if on circumference	cardioid
parabola	on directrix	strophoid
	foot of directrix	right strophoid
	reflection of focus in directrix	trisectrix of Maclaurin
	vertex	cissoid of Diocles
	focus	line
central conic	focus	circle
	centre	lemniscate of Bernoulli
Tschirnhausen's cubic	focus (of pedal)	parabola
cissoid	focus	cardioid
cardioid	cusp	Cayley's sextic
logarithmic spiral	pole	an equal spiral
involute of a circle	centre	Archimedes' spiral

Appendix III
Applications of Abel's Integral

The integral equation, occurring in nearly all of the ray-tracing solutions for non-uniform media, can be put in the general form

$$I = \int_{x_1}^{x_2} \frac{\kappa f(x) \mathrm{d}x}{[F^2(x) - \kappa^2]^{\frac{1}{2}}} \qquad \text{(III.1)}$$

Here I is a function of x with given values at the limits x_1 and x_2, and $f(x)$ is known. It is required to derive $F(x)$ subject to the boundary conditions of I. κ is a constant of an individual ray whose value can be determined from the value of I at x_1. The solution for a pencil of rays to have a given specified behaviour thus requires $F(x)$ to be independent of κ. For the source conditions used in this volume we have $f(x_1) = 1$ and x_2 will be a turning value, that is a value when I changes from a function increasing with x to one decreasing with x. At this value $F(x_2) = \kappa$, and the denominator becomes zero.

With these conditions the formal method is to make the substitutions

$$w = \int f(x)\mathrm{d}x; \qquad v = \kappa; \qquad u = F(x)$$

which converts the integral to Abel's form

$$I = \int_1^v \frac{v \dfrac{\mathrm{d}w}{\mathrm{d}u} \mathrm{d}u}{(u^2 - v^2)^{\frac{1}{2}}} \qquad \text{(III.2)}$$

Multiplication by $\mathrm{d}v/(v^2 - y^2)^{\frac{1}{2}}$ for an as yet unspecified function y, and integration in the range $[y, 1]$ gives

$$\int_y^1 \frac{v \mathrm{d}v}{(v^2 - y^2)^{\frac{1}{2}}} \int_1^v \frac{\dfrac{\mathrm{d}w}{\mathrm{d}u} \mathrm{d}u}{(u^2 - v^2)^{\frac{1}{2}}} = \int_y^1 \frac{I \mathrm{d}v}{(v^2 - y^2)^{\frac{1}{2}}}$$

Reversing the order of integration on the left-hand side and using the result

$$\int_y^1 \frac{v}{(u^2 - v^2)^{\frac{1}{2}}} \frac{\mathrm{d}v}{(v^2 - y^2)^{\frac{1}{2}}} = \frac{\pi}{2} \qquad \text{(III.3)}$$

gives

$$\int_y^1 \frac{\mathrm{d}w}{\mathrm{d}u} \mathrm{d}u \int_y^u \frac{v \mathrm{d}v}{(u^2 - v^2)^{\frac{1}{2}}(v^2 - y^2)^{\frac{1}{2}}} = \frac{\pi}{2} \int_y^1 \mathrm{d}w = \frac{\pi}{2} \int_y^1 f(x)\mathrm{d}x$$

and hence

$$\frac{\pi}{2}\int f(x)dx = \int_y^1 \frac{I\,dv}{(v^2 - y^2)^{\frac{1}{2}}} \tag{III.4}$$

III.A   For those situations where I is independent of v (or κ), such as the conditions governing specified refocusing of rays, the right-hand side is integrable, giving

$$\frac{\pi}{2}\int^x f(x)dx = I \log\left[\frac{1 + (1 - y^2)^{\frac{1}{2}}}{y}\right] \tag{III.5}$$

Finally putting $y = F(x)$ gives the required solutions. The limits are such that the upper limit in the integral on the left-hand side gives zero value.

(i) Cylindrical systems with repeated focusing (section 6.1)

$$f(x) = 1 \qquad F(x) = \eta(x) \qquad I = a \tag{III.6}$$

$$\eta(x) = \operatorname{sech}\frac{\pi x}{2a} \quad \text{(equation 6.3)}$$

(ii) Spherical systems (section 7.2)

$$f(r) = \frac{1}{r} \qquad F(r) = r\eta(r) \qquad I = 2\pi/y$$

$$\operatorname{sech}^{-1} r\eta(r) = -\frac{v}{2}\log r \tag{III.7}$$

or

$$r\eta(r) = \frac{2r^{v/2}}{r^v + 1}$$

the generalized fish-eye solutions (equations 7.11). $v = 2$ gives the original fish-eye.

(iii) Surface rays on the sphere (section 7.4)

$$f(\beta) = \sec\beta \qquad F(\beta) = \eta(\beta)\cos\beta \qquad I = \frac{2\pi}{v}$$

$$\operatorname{sech}^{-1}\left[\eta(\beta)\cos\beta\right] = v \log\left(\frac{1 + \sin\beta}{1 - \sin\beta}\right)^{\frac{1}{2}}$$

$$\cos\beta\eta(\beta) = \frac{2\cos^{v/2}\beta}{(1 + \sin\beta)^{\frac{1}{2}} + (1 - \sin\beta)^{\frac{1}{2}}} \quad \text{(equation 7.48)} \tag{III.8}$$

Also, the general result (equation 7.50) follows when $f(\beta)$ is replaced by

$$f'(\beta)/f(\beta)[1 - f^2(\beta)]^{\frac{1}{2}}$$

III.B When I is not independent of v (or therefore κ) integration depends upon the integrability of the function as given in equation III.4.

(iv) The Luneburg lens (section 7.2)

$$f(r) = \frac{1}{r} \qquad F(r) = r\eta(r) \qquad I = \frac{\pi}{2} - \frac{\sin^{-1} v}{2}$$

The right-hand side of equation III.4 now consists of two terms

$$\frac{\pi}{2} \log \left[\frac{1 + (1 - y^2)^{\frac{1}{2}}}{y} \right] - \frac{1}{2} \int_y^1 \frac{\sin^{-1} v}{(v^2 - y^2)^{\frac{1}{2}}} \, dv \qquad \text{(III.9)}$$

with the result

$$\log \left[\frac{1 + (1 - y^2)^{\frac{1}{2}}}{y} \right] - \tfrac{1}{2} \log \left[1 + (1 - y^2)^{\frac{1}{2}} \right] = -\log r \qquad \text{(III.10)}$$

or with $y = r\eta$; $\eta^2 = 2 - r^2$ (equation 7.14).

(v) Beam divider

These are generalizations of the Luneburg lens with $I = p\pi/4 - \sin^{-1} v/2 + \pi/2$, giving the result in (iv) modified to

$$(2 + p) \log \left[\frac{1 + (1 - y^2)^{\frac{1}{2}}}{y} \right] - \log \left[1 + (1 - y^2)^{\frac{1}{2}} \right] = -2 \log r$$

or

$$[1 + (1 - \eta^2 r^2)^{\frac{1}{2}}]^{p+1} = (r\eta)^{p+2} r^{-2} \quad \text{(equation 7.13)} \qquad \text{(III.11)}$$

(vi) Beam divider

$$I = q\pi/4 - \sin^{-1} v$$

and

$$[1 + (1 - \eta^2 r^2)^{\frac{1}{2}}]^{q-2} = (r\eta)^q r^{-2} \quad \text{(equation 7.17)} \qquad \text{(III.12)}$$

From the form of equation III.9 it can be seen that very few functions other than $\sin^{-1} v$ are integrable in the form needed to give the logarithmic function necessary for an algebraic solution. Hence for other focusing conditions, and in particular for two foci external to the spherical lens, the integral in equation III.9, and its eventual exponential, require computational methods and have been tabulated.

Appendix IV
The Radiation Patterns of Luneburg Lenses

The Luneburg lens forms an ideal transformer between a source with angular power distributions proportional to powers of cos α and radiation patterns in the far field proportional to a corresponding order of the lambda function.[50]

With the refractive-index law of equation 7.14

$$\eta = (2 - r^2)^{\frac{1}{2}}$$

and the rays have trajectories

$$x^2 - 2xy \cot \alpha + y^2(1 + 2 \cot^2 \alpha) = R^2 \qquad \text{(IV.1)}$$

Hence the ray intersects the sphere at the radial distance ρ from the axis given by

$$\rho = R \sin \alpha \qquad \text{(IV.2)}$$

Referring to Figure IV.1, the power radiated into the solid angle $d\Omega$ is converted into the power contained in the annulus $d\Omega$ by the relation

$$P_2(\rho) = P_1(\alpha) \sin \alpha d\alpha / \rho d\rho \qquad \text{(IV.3)}$$

From $\rho = R \sin \alpha$ we obtain $d\rho = R \cos \alpha d\alpha$, and since

$$\cos \alpha = (1 - \rho^2/R^2)^{\frac{1}{2}} \qquad \text{(IV.4)}$$

$P_1(\alpha)$ is converted to $P_1(\rho)$ through the relation $\sin \alpha = R/\rho$. Equation IV.3 becomes

$$P_2(\rho) = P_1(\alpha)/R^2 \cos \alpha = P_1(\rho)/R^2(1 - \rho^2/R^2)^{\frac{1}{2}} \qquad \text{(IV.5)}$$

For a circularly symmetrical aperture amplitude distribution $f(\rho)$ the far-field radiation pattern is given, in the Huygens–Kirchhoff approximation,[8 p.206] by the zero-order finite Hankel transform (normalized to unit radius)

$$g(u) = \int_0^1 f(\rho)J_0(u\rho)\rho d\rho \qquad u = \frac{2\pi R \sin \Theta}{\lambda} \qquad \text{(IV.6)}$$

λ being the wavelength of excitation, Θ the far field polar coordinate and constants and scaling factors have been omitted.

Since $f(\rho)$ is the *amplitude* distribution it is proportional to $[P_2(\rho)]^{\frac{1}{2}}$, with the

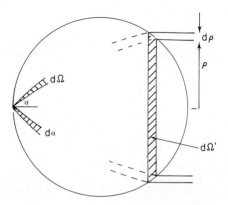

Figure IV.1 Luneburg lens transformer

result

$$g(u) = \int_0^1 \left[\frac{P_1(\rho)}{(1 - \rho^2)^{\frac{1}{2}}} \right]^{\frac{1}{2}} J_0(u\rho)\rho\,d\rho \qquad \text{(IV.7)}$$

There are several functions of $P_1(\rho)$ for which this integral is standard. A common description of the radiation from a microwave feed horn is to use a single function or a series of functions of the form

$$P_1(\alpha) = (\cos \alpha)^{2n+1} \rightarrow (1 - \rho^2)^{n+\frac{1}{2}}$$

Then $n = 0$ gives $P_1(\alpha) = \cos \alpha$ and

$$g(u) = \int_0^1 J_0(u\rho)\rho\,d\rho = \frac{J_1(u)}{u} \qquad \text{(IV.8)}$$

that is, a uniformly illuminated exit pupil and the diffraction-limited far-field pattern.

Generally

$$g(u) = \int_0^1 (1 - \rho^2)^{n/2} J_0(u\rho)\rho\,d\rho$$

$$= \frac{J_{n/2+1}(u)}{u^{n/2+1}} = \Lambda_{n/2+1}(u) \qquad \text{(IV.9)}$$

The result for $n = -1$ is of some interest. The primary pattern is $P_1(\alpha) = \sec \alpha$ and the far-field pattern is

$$g(u) = \int_0^1 \frac{J_0(u\rho)\rho\,d\rho}{(1 - \rho^2)^{\frac{1}{2}}} = \frac{\sin u}{u} \qquad \text{(IV.10)}$$

This is narrower than the diffraction limit and gives evidence of some super directivity. The same primary distribution applied to a paraboloid reflector gives uniform illumination in the aperture and hence the far-field pattern in equation IV.8.

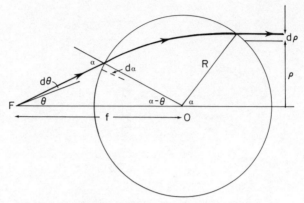

Figure IV.2 Luneburg lens with external source

Source patterns of the form $P_1(\rho) = [R_n(1 - 2\rho^2/R^2)]^2(1 - \rho^2)^{\frac{1}{2}}$, where $R_n(1 - 2\rho^2/R^2)$ is a circle polynomial of Zernike,[8] give radiation patterns

$$g(u) = \frac{J_{2n+1}(u)}{u}$$

Most source patterns can be generated by a series of such terms.

The identical result occurs if the source is not on the surface of the lens. If it is placed at a distance f from the *centre* of the lens (Figure IV.2), then the same analysis pertains with ρ above being given by

$$\rho = R \sin \alpha = f \sin \theta$$

with an adjustment when normalizing the integrals for the Hankel transform as in equation IV.6.

References

(1) (a) L. SILBERSTEIN 'Simplified method of tracing rays through any optical system of lenses, prisms or mirrors', Longmans Green, London 1918
(b) A. W. CONWAY and J. L. SYNGE (eds.) 'The collected papers of Sir William Rowan Hamilton—Vol. 1: Geometrical optics', Cambridge University Press, 1931, p. 10
(2) O. N. STAVROUDIS 'The optics of rays, wavefronts and caustics', Academic Press, 1972, pp. 97–102
(3) M. HERZBERGER 'Modern geometrical optics', Interscience Publishers, New York, 1958, p. 152
(4) J. D. EATON 'Zero phase fronts in microwave optics', *Trans. I.R.E.*, 1952, **AP-1**, 38
(5) O. N. STAVROUDIS 'Refraction of wavefronts: a special case', *J. Opt. Soc. Amer.*, 1969, **59**, 114
(6) J. D. LAWRENCE 'A catalog of special plane curves', Dover Publications, New York, 1972
(7) G. SALMON 'Higher plane curves', Chelsea Publishing, New York, 1960
(8) S. CORNBLEET 'Microwave optics', Academic Press, 1976, pp. 355, 370
(9) E. H. LINFOOT 'Recent advances in optics', Oxford University Press, 1955, p. 51
(10) J. C. MAXWELL 'On the description of oval curves and those having a plurality of foci', *Proc. Roy. Soc. Edinburgh*, April 1846, **II**
(11) E. CUNNINGHAM 'The principle of relativity and an extension thereof', *Proc. London Math. Soc.* (2), 1909, **8**, 77
(12) H. BATEMAN 'The transformation of coordinates which can be used to transform one physical problem into another', *Proc. London Math. Soc.* (2), 1910, **8**, 469
(13) (a) J. B. SCARBOROUGH 'The caustic curve of an off-axis parabola', *Applied Optics*, 1964, **3**, 1445
(b) H. J. STALZER 'Comment on the caustic curve of a parabola', *Applied Optics*, 1965, **4**, 1205
(14) S. CORNBLEET 'Feed position for the parabolic reflector with offset pattern', *Electron. Lett.*, 1979, **15**, 211
(15) M. BORN and E. WOLF 'Principles of optics', Pergamon Press, 1959, p. 246
(16) R. DAMIEN 'Théorème sur les surfaces d'onde en optique géometrique', Gauthier Villars, Paris, 1955
(17) S. CORNBLEET 'New geometrical method for the design of optical systems', *Proc. I.E.E.* (Pt. H: *Microwaves, Optics and Acoustics*), 1979, **3**, p. 78
(18) S. CORNBLEET and B. J. GUNNEY 'Corrected plane reflectors', I.E.E. Conf. Publ. 169 (Pt. 1: Antennas and Propagation), 1978, p. 201
(19) S. ROBERTS 'Historical note on Dr. Graves' theorem on confocal conics', *Proc. London Math. Soc.*, 1881, **12**, 120
(20) J. L. SYNGE 'Geometrical optics', Cambridge University Press, 1937
(21) H. A. ATWATER 'Introduction to general relativity', Pergamon Press, 1974, p. 69
(22) K. G. BUDDEN 'Radio waves in the ionosphere', Cambridge University Press, 1961
(23) M. BORN and E. WOLF 'Principles of optics', Pergamon Press, 1959, p. 674
(24) J. BROWN 'Microwave lenses', Methuen, 1953
(25) R. K. LUNEBURG 'The mathematical theory of optics', University of California Press, 1964
(26) J. BROWN 'Microwave lenses', Methuen, 1953, p. 89

148

(27) M. HERZBERGER 'Modern geometrical optics', Interscience Publishers, New York, 1958, p. 30
(28) R. W. WOOD 'Physical optics', Macmillan, 1905, p. 72
(29) P. F. BYRD and M. D. FRIEDMAN 'Handbook of elliptic integrals for engineers and scientists', Springer Verlag, 1971
(30) S. EXNER 'Die Physiologie der facettirten Augen von Krebsen und Insekten', Deutike, Leipzig and Wien, 1891
(31) W. LENZ 'Theory of optical images' in 'Probleme der modernen Physik' (ed. P. Debye), Hirsel Press, Leipzig, 1928, p. 198
(32) A. S. GUTMAN 'Modified Luneburg lens', J. Appl. Phys., 1954, 23, 855
(33) J. E. EATON 'On spherically symmetric lenses', Trans. I.R.E., 1952, AP-4, 66
(34) G. TORALDO DI FRANCIA 'A family of perfect configuration lenses of revolution', Optica Acta, 1955, 1, 157
(35) S. CORNBLEET 'Microwave optics', Academic Press, 1976, p. 138
(36) J. C. MAXWELL 'On the condition that, in the transformation of any figure . . . in three dimensions, every angle in the new figure shall be equal to the corresponding angle in the original figure', Proc. London Math. Soc., 1871–73, 4, 117
(37) S. CORNBLEET and M. C. JONES 'The transformation of spherical nonuniform lenses' Proc. I.E.E. (Pt. H), 1982, 129, 321
(38) P. MOON and D. E. SPENCER 'Field theory for engineers', Van Nostrand, 1961, p. 348
(39) (a) S. P. MYERS 'Parallel plate optics for rapid scanning', J. Appl. Phys., 1947, 18, 221
    (b) R. F. RINEHART 'A family of designs for rapid scanning radar antennas', Proc. I.R.E., June 1952, 686
(40) K. S. KUNZ 'Propagation of microwaves between a pair of doubly curved conducting surfaces', J. Appl. Phys., 1954, 25, 642
(41) S. CORNBLEET and P. J. RINOUS 'Generalised formulas for equivalent geodesic and nonuniform lenses', Proc. I.E.E. (Pt. H), 1981, 128, 95
(42) (a) D. KASSAI and E. MAROM 'Aberration corrected rounded edge geodesic lenses', J. Opt. Soc. Amer., 1979, 69, 1242
    (b) W. H. SOUTHWELL 'Geodesic optical waveguide analysis', J. Opt. Soc. Amer., 1977, 67, 1293
    (c) S. SOTTINI, V. RUSSO and G. C. RIGHINI 'General solution of the problem of perfect geodesic lenses for integrated optics', J. Opt. Soc. Amer., 1979, 69, 1248
(43) H. GOLDSTEIN 'Classical mechanics', Addison Wesley, 1964
(44) (a) H. A. BUCHDAHL 'Conformal transformations and conformal invariance of optical systems', Optik, 1975, 43, 259
    (b) H. A. BUCHDAHL 'Kepler problem and Maxwell fish-eye', Amer. J. Phys., 1978, 46, 840
    (c) H. A. BUCHDAHL 'Luneburg lens: unitary invariance and point characteristic', J. Opt. Soc. Amer., 1983, 73, 490
(45) P. COLLAS 'Equivalent potentials in classical mechanics', J. Math. Phys., 1981, 22, 2512
(46) E. KASNER 'The trajectories of dynamics', Trans. Amer. Math. Soc., 1906, 7, 401
(47) H. BATEMAN 'The conformal transformations of a space of four dimensions and their applications to geometrical optics', Proc. London Math. Soc., 1908, 7, 70
(48) (a) D. WEINGARTEN 'Complex symmetries of electrodynamics', Annals Phys., 1973, 76, 510
    (b) K. IMAEDA 'A new formulation of classical electrodynamics', Nuovo Cimento, 1976, 32B, 138
(49) S. CORNBLEET 'Geometrical optics reviewed: a new light on an old subject', Proc. I.E.E.E., 1983, 71, 471
(50) E. JAHNKE and F. EMDE 'Tables of functions', Dover Publications, 1945, p. 180

Index

149